Coastal Fish Research

Coastal Fish Research

Editor

Francesco Tiralongo

MDPI • Basel • Beijing • Wuhan • Barcelona • Belgrade • Manchester • Tokyo • Cluj • Tianjin

Editor
Francesco Tiralongo
Department of Biological,
Geological and Environmental
Sciences
University of Catania
Catania
Italy

Editorial Office
MDPI
St. Alban-Anlage 66
4052 Basel, Switzerland

This is a reprint of articles from the Special Issue published online in the open access journal *Journal of Marine Science and Engineering* (ISSN 2077-1312) (available at: www.mdpi.com/journal/jmse/special_issues/coastal_fish_research).

For citation purposes, cite each article independently as indicated on the article page online and as indicated below:

LastName, A.A.; LastName, B.B.; LastName, C.C. Article Title. *Journal Name* **Year**, *Volume Number*, Page Range.

ISBN 978-3-0365-1450-5 (Hbk)
ISBN 978-3-0365-1449-9 (PDF)

Contents

About the Editor

Francesco Tiralongo

Francesco Tiralongo is an ichthyologist at the University of Catania and has authored numerous international peer-reviewed scientific articles and conferences on Mediterranean fishes. He is an avid explorer of coastal aquatic habitats and is mainly interested in the study of coastal fish species. His current research focuses on coastal fish biology and ecology, fisheries, and non-indigenous species in the Mediterranean Sea.

Journal of
Marine Science and Engineering

MDPI

Editorial

Coastal Fish Research

Francesco Tiralongo

Department of Biological, Geological and Environmental Sciences, University of Catania, 95124 Catania, Italy; francesco.tiralongo@unict.it

Coastal fish are key components of marine ecosystems, influencing, directly or indirectly, marine life worldwide. Furthermore, among coastal fish, there are many species that represent target species for many fisheries (commercial and recreational) and contribute considerably to the economy of several coastal countries. Despite this, the biology and ecology of several species still remain little known or unknown, with several species waiting to be described. Furthermore, many fish stocks are being overexploited and in urgent need of sustainable management. Another interesting side of coastal fish is their importance as bioindicators of the status of the aquatic ecosystems. Indeed, often, coastal fish are directly exposed (and sometimes seriously threatened) to a variety of human impacts (e.g., pollution, habitat destruction, overfishing) [1]. On the other hand, invasive alien coastal fish represent in many areas a serious threat to marine biodiversity and in some cases to economy and human health.

It thus appears clear that understanding and deepening knowledge around biology and ecology of coastal fish is one of the main challenges for marine biologists and ecologists. For example, in an era in which invasive alien species threaten marine biodiversity, but can also have economic and human health impacts, the understanding of the dynamics and processes underlying biological invasions is of fundamental importance for effective management and conservation of biodiversity and ecological integrity [2]. Furthermore, the increasing marine pollution represents a serious threat to coastal fish and to marine life in general, although this phenomenon encompasses a great variety of aspects and effects on marine fish are often difficult to detect and quantify. However, some types of laboratory analysis on coastal fish fauna can give good indications of the general status of the marine environment and of the effects that specific pollutants can have on fish populations.

This special edition is intended to be a contribution to the knowledge of several aspects of fish biology and ecology: coastal fish diversity, coastal fisheries and commercial species, biological invasions process and non-indigenous species, interactions between fish and their environment, ethology, bioaccumulation and reproduction [1–12]. I hope this special issue will be helpful for all marine biologists involved in studies on fish biology and ecology.

Funding: This research received no external funding.

Conflicts of Interest: The author declares no conflict of interest.

Citation: Tiralongo, F. Coastal Fish Research. *J. Mar. Sci. Eng.* **2021**, *9*, 546. https://doi.org/10.3390/jmse9050546

Received: 12 May 2021
Accepted: 17 May 2021
Published: 18 May 2021

References

1. Salvaggio, A.; Pecoraro, R.; Copat, C.; Ferrante, M.; Grasso, A.; Scalisi, E.M.; Ignoto, S.; Bonaccorsi, V.S.; Messina, G.; Lombardo, B.M.; et al. Bioaccumulation of Metals/Metalloids and Histological and Immunohistochemical Changes in the Tissue of the European Hake, *Merluccius merluccius* (Linnaeus, 1758) (Pisces: Gadiformes: Merlucciidae), for Environmental Pollution Assessment. *J. Mar. Sci. Eng.* **2020**, *8*, 712. [CrossRef]
2. Tiralongo, F.; Messina, G.; Lombardo, B. Invasive Species Control: Predation on the Alien Crab *Percnon gibbesi* (H. Milne Edwards, 1853) (Malacostraca: Percnidae) by the Rock Goby, *Gobius paganellus* Linnaeus, 1758 (Actinopterygii: Gobiidae). *J. Mar. Sci. Eng.* **2021**, *9*, 393. [CrossRef]
3. Macali, A.; Semenov, A.; De Mendoza, F.P.; Dinoi, A.; Bergami, E.; Tiralongo, F. Relative Influence of Environmental Factors on Biodiversity and Behavioural Traits of a Rare Mesopelagic Fish, *Trachipterus trachypterus* (Gmelin, 1789), in a Continental Shelf Front of the Mediterranean Sea. *J. Mar. Sci. Eng.* **2020**, *8*, 581. [CrossRef]
4. Tiralongo, F.; Messina, G.; Lombardo, B.M. Biological Aspects of Juveniles of the Common Stingray, *Dasyatis pastinaca* (Linnaeus, 1758) (Elasmobranchii, Dasyatidae), from the Central Mediterranean Sea. *J. Mar. Sci. Eng.* **2020**, *8*, 269. [CrossRef]
5. Candelma, M.; Marisaldi, L.; Bertotto, D.; Radaelli, G.; Gioacchini, G.; Santojanni, A.; Colella, S.; Carnevali, O. Aspects of Reproductive Biology of the European Hake (*Merluccius merluccius*) in the Northern and Central Adriatic Sea (GSA 17-Central Mediterranean Sea). *J. Mar. Sci. Eng.* **2021**, *9*, 389. [CrossRef]
6. Bruneel, S.; Van Echelpoel, W.; Ho, L.; Raat, H.; Schoeters, A.; De Troyer, N.; Sor, R.; Ponton-Cevallos, J.; Vandeputte, R.; Van der Heyden, C.; et al. Assessing the Drivers behind the Structure and Diversity of Fish Assemblages Associated with Rocky Shores in the Galapagos Archipelago. *J. Mar. Sci. Eng.* **2021**, *9*, 375. [CrossRef]
7. Di Blasi, D.; Canese, S.; Carlig, E.; Parker, S.; Pisano, E.; Vacchi, M.; Ghigliotti, L. The Challenge to Observe Antarctic Toothfish (*Dissostichus mawsoni*) under Fast Ice. *J. Mar. Sci. Eng.* **2021**, *9*, 255. [CrossRef]
8. Čekovská, K.; Šanda, R.; Eliášová, K.; Kovačić, M.; Zogaris, S.; Pappalardo, A.M.; Soukupová, T.; Vukić, J. Population Genetic Diversity of Two Marine Gobies (Gobiiformes: Gobiidae) from the North-Eastern Atlantic and the Mediterranean Sea. *J. Mar. Sci. Eng.* **2020**, *8*, 792. [CrossRef]
9. Lu, Y.; Yu, J.; Lin, Z.; Chen, P. Environmental Influence on the Spatiotemporal Variability of Spawning Grounds in the Western Guangdong Waters, South China Sea. *J. Mar. Sci. Eng.* **2020**, *8*, 607. [CrossRef]
10. Park, J.M.; Riedel, R.; Ju, H.H.; Choi, H.C. Fish Assemblage Structure Comparison between Freshwater and Estuarine Habitats in the Lower Nakdong River, South Korea. *J. Mar. Sci. Eng.* **2020**, *8*, 496. [CrossRef]
11. Spyridopoulou, R.N.A.; Langeneck, J.; Bouziotis, D.; Giovos, I.; Kleitou, P.; Kalogirou, S. Filling the Gap of Data-Limited Fish Species in the Eastern Mediterranean Sea: A Contribution by Citizen Science. *J. Mar. Sci. Eng.* **2020**, *8*, 107. [CrossRef]
12. Zhang, J.; Ren, C.; Zhang, H.; Yin, F.; Zhang, S.; Wan, R.; Kitazawa, D. Review of Estimating Trophic Relationships by Quantitative Fatty Acid Signature Analysis. *J. Mar. Sci. Eng.* **2020**, *8*, 1030. [CrossRef]

Journal of
Marine Science and Engineering

MDPI

Article

Invasive Species Control: Predation on the Alien Crab *Percnon gibbesi* (H. Milne Edwards, 1853) (Malacostraca: Percnidae) by the Rock Goby, *Gobius paganellus* Linnaeus, 1758 (Actinopterygii: Gobiidae)

Francesco Tiralongo [1,2,*], Giuseppina Messina [1] and Bianca Maria Lombardo [1]

1 Department of Biological, Geological and Environmental Sciences, University of Catania, 95124 Catania, Italy; giuseppina.messina@unict.it (G.M.); bm.lombardo@unict.it (B.M.L.)
2 Ente Fauna Marina Mediterranea, 96012 Avola, Italy
* Correspondence: francesco.tiralongo@unict.it

Abstract: Invasive alien species (IAS) are one of the greatest causes of native species extinction. Indeed, they represent a global threat for biodiversity and can also affect the economy and human health. The colonization success of IAS is presumably not only due to their biological and ecological characteristics, but also to the lack of predators and/or parasites in the invaded new areas. In the present work, we demonstrate evidence of predation of the invasive alien crab *Percnon gibbesi* (H. Milne Edwards, 1853) by the Rock Goby *Gobius paganellus* Linnaeus, 1758. The diet of *G. paganellus* was studied analyzing the stomach content of 162 specimens collected in the central Mediterranean Sea. The results obtained from the calculation of the diet indices, namely, frequency of occurrence (%F), percentage weight (%W), percentage abundance (%N), and the Index of Relative Importance (%IRI), showed that small benthic crustaceans were the main prey types. Additionally, these indices and the Levins' index (B*i*) clearly indicated that the invasive crab *P. gibbesi* was by far the most abundant prey type in the diet of *G. paganellus*. The relevance of this predator–prey interaction and the role of native species for the biological control of invasive ones are discussed. We also provide a general view on the diet of *G. paganellus* and other biological and ecological aspects of specimens studied from the central Mediterranean Sea.

Keywords: Mediterranean Sea; invasive species; non-indigenous species; biological control; prey–predator interactions

Citation: Tiralongo, F.; Messina, G.; Lombardo, B.M. Invasive Species Control: Predation on the Alien Crab *Percnon gibbesi* (H. Milne Edwards, 1853) (Malacostraca: Percnidae) by the Rock Goby, *Gobius paganellus* Linnaeus, 1758 (Actinopterygii: Gobiidae). *J. Mar. Sci. Eng.* **2021**, *9*, 393. https://doi.org/10.3390/jmse9040393

Academic Editor: Gualtiero Basilone

Received: 22 February 2021
Accepted: 2 April 2021
Published: 7 April 2021

1. Introduction

Non-indigenous species (NIS), and in particular invasive alien species (IAS), represent a serious threat to ecosystems' integrity, interfering with key ecological processes. Several studies have demonstrated that IAS can reduce the abundance and presence of native species through predation or competition and can alter food webs and community structure [1,2]. Indeed, they are considered to be the greatest cause of native species extinctions after habitat destruction [3]. In some cases, these alterations to the ecosystem can cause severe economic losses, sometimes exceeding those of natural disasters, and represent a threat to human health [4,5]. Although the dynamics of biological invasions are complex and often unclear, the results of several studies have suggested that areas with high species richness are more resistant to biological invasions than areas poor in species. In the former areas, the scarce availability of a free ecological niche represents an obstacle to the spread of IAS, reducing the possibility of settlement and/or expansion. This hypothesis is known as the "biotic resistance hypothesis" [6,7]. In this context, but also in general, the role of native predators can be relevant for the biological control of IAS. Indeed, in some cases, native species can have a high potential to exert predatory control on invasive ones [8]. Among these, coastal fish species can potentially play a key role in the predation of IAS.

Gobies comprise more than 1900 species, and are the largest family (Gobiidae) of marine fishes [9,10]. In the Mediterranean Sea, this family includes 73 species, thus being the most diverse fish group of the region [11]. Despite this and their relevance for marine ecosystems, an increasing interest in gobies only started recently, with several published studies on their biology and ecology and the description of new species [12–18].

Gobius paganellus Linnaeus, 1758 is a medium-sized goby (Gobiidae), whose distribution extends from the Mediterranean and Black Sea to the eastern Atlantic (from Scotland to Senegal), including the Atlantic islands of the Azores, Madeira, and the Canaries [19]. Furthermore, it represents one of the few fish species that has performed an anti-Lessepsian migration, reaching the Red Sea through the Suez Canal [20]. The maximum reported total length (TL) is 14 cm for males and 14.3 cm for females [12], but common sizes range from about 8 to 12 cm TL, and no marked differences in size are present between sexes [12,21,22]. It is a shallow waters species generally common on rocky bottoms, often occurring in tidal pools [22,23]. Total length at first maturity was estimated at 5.2–11.4 cm TL, depending on the sex and location, a size that can be reached at 1–3 years of age [12,21,22,24,25]. Crustaceans are the main prey types in the diet of *G. paganellus* [12,21,24,26], and, depending on the size, area, and season, there is a variation in the dominance of the different species on which this goby feeds. Hence, *G. paganellus* is an opportunistic predator. However, among crustaceans, amphipods and isopods, followed by Caridea and Brachyura, were recorded to be the most common prey in the diet of the species [12,21,24,26].

Percnon gibbesi (H. Milne Edwards, 1853), commonly known as the Nimble Spray Crab, is a medium-sized crab whose distribution extends from California to Chile in the Pacific Ocean, from North Carolina to Brazil in the Western Atlantic Ocean, and from the Azores to Angola in the Eastern Atlantic Ocean [27]. After its first record in 1999 in the Mediterranean Sea, *P. gibbesi* has undergone a rapid expansion in the whole basin [28,29]. By year 2000, the rapid spread of the species was recorded in Sicily [30]. The great colonization success of this crab was attributed to several factors, such as an increase of sea water temperatures, the absence of competitors, the availability of unoccupied ecological niches, and life history characteristics such as some aspects of its reproductive biology and long planktotrophic larval stage [31]. However, to the best of our knowledge, the predation of this species by a native species has never been demonstrated for the Mediterranean Sea, because *P. gibbesi* was never found in the stomach of any analyzed species and no predators were directly observed to prey on the crab (e.g., [7,12,15,21]). Yet, native species can prey on invasive ones, and, in some cases, contribute to controlling their populations [8,32–36]. Although several means of introduction have been hypothesized for *P. gibbesi* (i.e., aquarium release, larval drift through the Strait of Gibraltar, adult migration, ballast waters, ship hulls), the most likely appears to be the introduction through shipping [29,37,38]. This species was often found among boulders covered by algal mats or in which this covering was almost absent, in very shallow waters [29]. This habitat simultaneously provides protection from predators and food supply, represented by small algae and other small sessile organisms [30]. *Percnon gibbesi* has been considered mainly herbivorous (algae-eating), but some authors demonstrated its opportunistic feeding behavior [29,30]. Indeed, the species was observed to prey on hermit crabs and polychaetes, and analysis of stomach content revealed that gastropods and crustaceans can constitute a considerable part of its diet [39,40]. In order to avoid diurnal predators, *P. gibbesi* is more active starting from the evening hours [31,40]. In Mediterranean waters, ovigerous females were observed between May and September. Sexual dimorphism is present, with males showing chela length with a larger positive allometry than that of females [31]. According to some authors, *P. gibbesi* is a potential competitor of *Pachygrapsus marmoratus* (J.C. Fabricius, 1787), both for space and food resources, and a similar situation could be present with *Eriphia verrucosa* (Forskål, 1775), although both native crabs usually prefer inhabiting the rocky intertidal and very shallow waters, among crevices and holes or under boulders [31].

In the present study, we provide the first direct evidence of substantial predation of a native species, *G. paganellus*, on the invasive alien crab, *P. gibbesi*, emphasizing the role of the goby in the biological control of this invasive species. Furthermore, we also provide the first recorded data on some aspects of the biology and ecology of *G. paganellus* from the Ionian Sea (central Mediterranean Sea): length–weight relationship, total length–standard length relationship, size frequency distribution, diet, and feeding habits. For these latter two points, we analyzed and discussed the prey selectivity of the species.

2. Materials and Methods

A total of 162 specimens of *G. paganellus* were collected in the Ionian Sea (south-east coast of Sicily, central Mediterranean Sea), in an area between the localities of Avola and Noto, along a coastline extending for about 2 km (Figure 1). In order to better represent the investigated population, two collection areas were selected, that of Noto represented by a small semi natural harbor (36.86994 N, 15.13738 E) with rocky bottom and boulders and that of Avola represented by a natural rocky bottom area (36.88628 N, 15.14041 E). These areas were distant from each other by about 2 km and had a depth range of 0.1–1.5 m. Sampling was performed during daytime with calm sea conditions, between 4 September and 5 December 2020. In this period, a total of 12 samplings were carried out, with an average of 13.5 specimens per sampling day (Table 1).

The specimens were caught with a fishing rod, using as bait fresh pieces of the Deep-water Rose Shrimp, *Parapenaeus longirostris* (H. Lucas, 1846). Specimens were killed in ice water immediately after their capture. Subsequently, each specimen was weighed and measured (total length and standard length). Total length and weight measures were used for the length–weight relationship following the formula: $W = aTL^b$, where W is the weight in grams (g), TL is the total length in centimetres (cm), a is the intercept, and b is the slope of the regression curve; when $b = 3$, the increase in weight is isometric; contrariwise, when the value of b is different to 3, the weight increase is allometric. In this latter case, if $b > 3$, the weight increase is positive allometric, and if $b < 3$, the weight increase is negative allometric. A one-sample t-test was used to verify the null hypothesis of the isometric growth (H_0: $b = 3$). Length–weight relationships are valuable measurements in order to evaluate and compare the health conditions of species' population [41,42]. Total length (TL) and standard length (SL) measures were used for the total length–standard length relationship following the formula: $TL = \alpha SL + \beta$, where α is the slope and β is the intercept of the regression line.

Table 1. Means of the basic measurements per sampling day for *Gobius paganellus*; N = total number of specimens; N Avola = specimens collected at Avola; N Noto = specimens collected at Noto; N Prey = specimens with at least one prey in their stomach; TL = total length; SL = standard length; W = weight.

Date	N	N Avola	N Noto	N Prey	TL (cm)	SL (cm)	W (g)
4 September 2020	15	5	10	11	9.52	7.89	11.45
14 September 2020	11	2	9	7	9.24	7.68	10.52
21 September 2020	12	4	8	9	9.54	8	11.26
28 September 2020	16	8	8	12	9.50	7.99	11.78
2 October 2020	12	5	7	10	9.82	8.20	12.91
9 October 2020	15	10	5	13	9.83	8.25	12.09
16 October 2020	13	3	10	9	9.09	7.57	10.05
27 October 2020	14	6	8	6	9.56	7.96	11.80
3 November 2020	12	8	4	8	9.59	8.06	11.48
10 November 2020	15	8	7	10	9.19	7.71	10.69
22 November 2020	14	9	5	11	9.8	8.19	12.85
5 December 2020	13	4	9	9	9.49	7.95	11.26

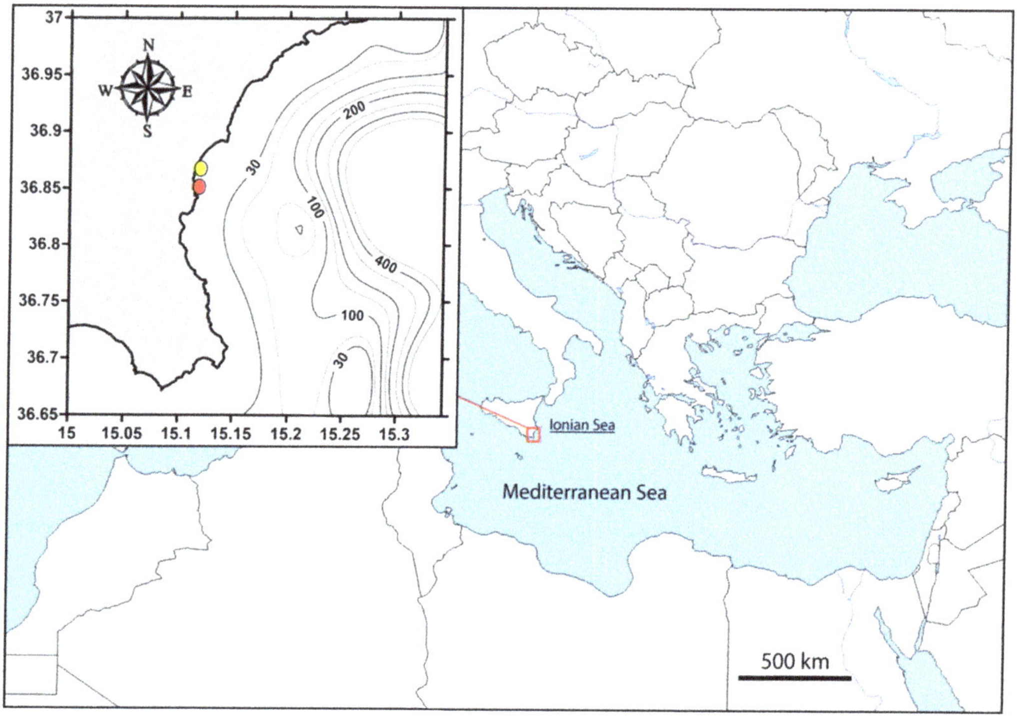

Figure 1. Study areas of Noto (**in red**) and Avola (**in yellow**) in the Ionian coast of Sicily (central Mediterranean Sea).

After measurements, the stomach was removed and the content analyzed under a stereoscopic microscope. All the prey items found were counted, washed in clean seawater and dried with blotter paper, identified to the lowest taxonomic level possible using identification manuals [43,44], and weighed to the nearest 0.001 g. After examination, stomach contents and gobies were preserved in ethanol.

The frequency of occurrence—%F (defined as the number of stomachs in which each specific prey type is represented expressed as percentage), percentage weight—%W (defined as the weight of each prey type in the stomachs examined expressed as percentage), percentage abundance—%N (defined as the number of individuals of each prey type in the stomachs examined expressed as percentage), and the Index of Relative Importance—%IRI (calculated by summing %N and %W values and multiplying with %F value) were calculated for each taxon [45,46]. The percentage of empty stomachs (vacuity index)—V% (number of empty stomachs/total number of stomachs examined) × 100, was also calculated.

Prey were grouped into three categories on the basis of their percentage abundance (%N) [47]: dominant (N > 50%), secondary (10% < N < 50%), and accidental (N < 10%).

Standardized Levins' index (B*i*) was used to evaluate the breadth of the diet [48]:

$$B = \frac{1}{\sum p_j^2} \quad (1)$$

$$Bi = \frac{B - 1}{B_{max} - 1} \quad (2)$$

where p_j is the relative frequency specimens in the jth prey item and B_{max} is the total number of prey item categories found. B*i* is comprised between 0 and 1. The higher the value of this index, the wider the trophic niche of the species will be. Hence, if $Bi < 0.40$, the species is considered a "specialist feeder", if 0.40 < Bi < 0.60 the species is considered an "intermediate feeder", and if B*i* > 0.60, the species is considered a "generalist feeder" [49].

A cumulative prey curve [50] was computed with R Studio [51] using the "vegan" package in order to evaluate whether the number of analyzed stomachs was sufficient to describe the diet of the species. The estimated number of prey groups with the associated SD were plotted against the cumulative number of individuals whose stomach was examined.

3. Results

One hundred and fifteen out of a total of 162 specimens examined of *G. paganellus* had prey in their stomachs; thus, the vacuity index (percentage of empty stomachs) was V% = 29.01. The cumulative prey curve approached an asymptote, suggesting that the analysis of about 100 not empty stomachs provides a reliable description of the diet of the goby (Figure 2). Analysis of stomach contents of *G. paganellus* included 20 prey types. Data showed that *G. paganellus* feed mainly on small benthic crustaceans (Table 2). In particular, values of %F, %W, %N, and %IRI indicated that *P. gibbesi*, which accounted for about one third of all prey items, was by far the most abundant prey type. This was also strongly supported by the Levins' index value (B*i* = 0.39), indicating a narrow trophic niche ("specialist feeder"). The carapace width (CW) of the specimens of *P. gibbesi* found in the stomachs was about or less than 1 cm. On a total of 20 prey types recorded, no "dominant" (N > 50%) prey types were found, and the only "secondary" prey type (10% < N < 50%) was represented by *P. gibbesi*. Hence, except for *P. gibbesi*, all the prey types fall into the category "accidental preys" (N < 10%) (Table 2). However, among these latter preys, the most represented were some Brachyura (*Pisidia* sp., *Xantho* sp., and juvenile specimens of *P. marmoratus*) and the hermit crab *Clibanarius erythropus* (Latreille, 1818). *Eriphia verrucosa* was the less represented crab (%F = 4.35). The other groups of crustaceans found in the stomachs of the Rock Goby were Amphipoda, Caridea, and Isopoda. These latter groups of crustaceans and Algae showed similar values in diet indices (e.g., %F = 6.09–6.96). Mollusca (gastropods) were overall less represented (%F = 0.87–1.74, excluding value for not identified Gastropoda), followed by Polychaeta and Seagrass (%F = 2.61 and 0.87, respectively) (Table 2).

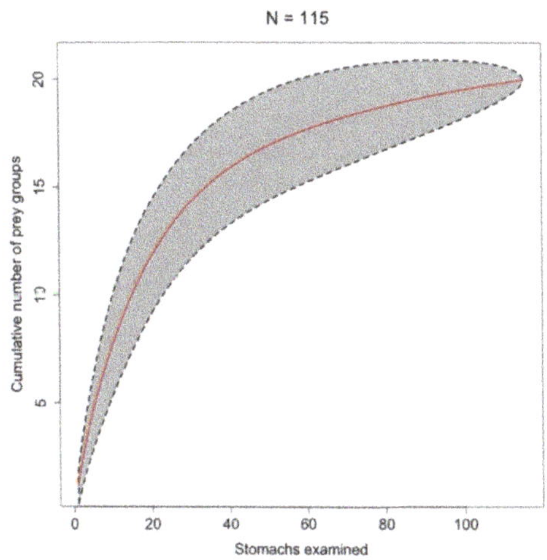

Figure 2. Cumulative prey curve (**in red**) as a function of sample size for all stomachs analyzed of *Gobius paganellus*. Standard deviation (SD) in grey delimited by dashed black lines.

Table 2. Diet composition of *G. paganellus* (N = 115) from the Ionian Sea (central Mediterranean Sea). %F = percentage frequency of occurrence; %N = percentage in number; %W = percentage in biomass; IRI = index of relative importance of prey items and its percentage (%IRI). Results for *Percnon gibbesi* are underlined.

Prey types	%F	%N	%W	IRI	%IRI
Mollusca	**11.31**	**9.36**	**10.24**	**49.916**	**1.62**
Aplysia sp.	0.87	0.72	1.63	2.0435	0.07
Bittium lacteum	0.87	0.72	0.54	1.0956	0.04
Patella caerulea	1.74	1.44	1.01	4.2610	0.14
Tritia cuvierii	1.74	1.44	1.90	5.8116	0.19
Gastropoda n.i.	4.35	3.60	3.67	31.6087	1.02
Trochidae n.i.	1.74	1.44	1.49	5.0956	0.16
Polychaeta	**2.61**	**2.16**	**0.95**	**8.113**	**0.26**
Polychaetes n.i.	2.61	2.16	0.95	8.113	0.26
Crustacea	**89.59**	**77.73**	**79.65**	**2925.5660**	**94.42**
Clibanarius erythropus	6.96	6.47	4.27	74.7130	2.41
Eriphia verrucosa	4.35	3.60	3.49	30.8415	1.00
Pachygrapsus marmoratus	5.22	5.04	5.03	52.5391	1.70
<u>*Percnon gibbesi*</u>	<u>33.91</u>	<u>30.22</u>	<u>40.15</u>	<u>2386.4608</u>	<u>77.01</u>
Pisidia sp.	6.09	5.04	12.42	106.2782	3.43
Xantho sp.	6.96	5.76	4.17	69.0783	2.23
Amphipoda n.i.	6.09	5.04	1.22	38.1234	1.23
Brachyura n.i.	6.96	5.76	4.61	69.3753	2.24
Caridea n.i.	6.96	5.76	1.49	50.4347	1.63
Isopoda n.i.	6.09	5.04	2.80	47.7217	1.54
Algae	**6.09**	**5.04**	**3.53**	**52.1652**	**1.68**
Seagrass (*Posidonia oceanica*)	0.87	0.72	0.43	1.0005	0.03
Digested	**6.09**	**5.04**	**5.19**	**62.2695**	**2.01**

In the specimens sampled in the small semi natural harbor (N = 90), *P. gibbesi* was found in 43.3% (N = 39) of the stomachs. In contrast, *P. gibbesi* was absent from the stomach contents of specimens sampled in the nearby natural rocky bottom area. In the whole sample of *G. paganellus* with at least one prey in their stomach (N = 115), *P. gibbesi* was found in 33.9% of the stomachs analyzed. However, in order to compare the diet of the goby in the aforementioned locations, removing *P. gibbesi* from the analysis did not find any significant differences between them. In all cases, crabs and hermit crabs represented the most common prey types of the species, although no clear dominance was evident for any prey type (with the exception of *P. gibbesi*).

The size frequency distribution of the sampled specimens (N = 162) showed a size range of 7.1–11.6 cm TL, with a mean of 9.5 cm (Table 3). The size frequency distribution was multimodal (Figure 3). The total length–weight relationship showed an isometric growth (b = 2.959; *p*-value > 0.05) (Table 3), while the parameters of the linear regression of total length–standard length relationship were: TL = 1.136SL + 0.4822 (Figure 4). Means of the basic measurements per sampling day are reported in Table 1.

Table 3. Total length, standard length, weight, and total length–weight relationship parameters of *G. paganellus* in the Ionian Sea (central Mediterranean Sea); a = intercept of the regression curve; b = slope; r^2 = coefficient of correlation; C.I. = 95% confidence interval.

Total Length (TL)-cm			Standard Length (SL)-cm			Weight (W)-g		
Range	**Mean**	**SD**	**Range**	**Mean**	**SD**	**Range**	**Mean**	**SD**
7.1–11.6	9.5	1.11	5.9–9.6	7.9	0.97	4.98–20.14	11.53	3.80
a	C.I. a		b	C.I. b		r^2	*p*-value	
0.014	0.011–0.018		2.959	2.849–3.068		0.95	>0.05	

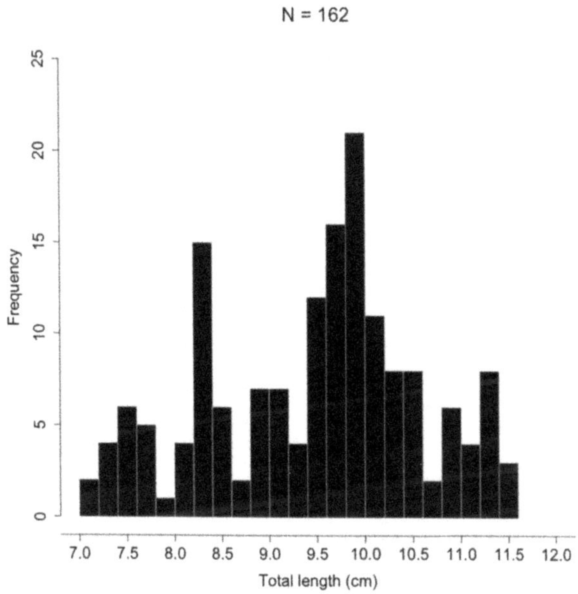

Figure 3. Size distribution (TL in cm) of sampled specimens of *G. paganellus*.

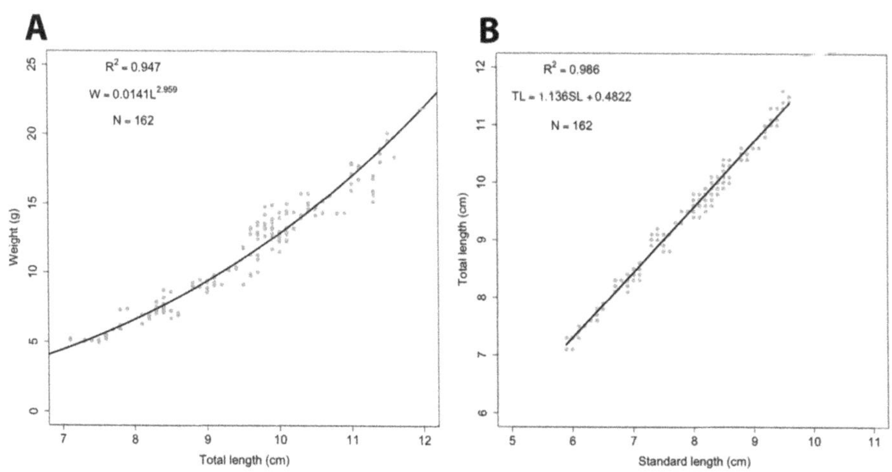

Figure 4. Total length–weight relationship (**A**) and total length–standard length relationship (**B**) of *G. paganellus*.

4. Discussion

The diet analysis of *G. paganellus* clearly showed the dominance of the alien crab *P. gibbesi*. Indeed, this species of crab was found in the stomach of a considerable percentage of specimens (%F = 33.91; %IRI = 77.01). Considering the small size of the crabs found in the stomachs, they were all juveniles. A noteworthy fact is that *P. gibbesi* was found only in the stomach of specimens sampled in the small harbor (N = 90). This can be explained by the fact that gobies are usually generalists and opportunistic feeders that take advantage of the most available prey. Furthermore, the gobies' diet was found to be often dependent on area and season [14,15,52]. Based on this, the relative abundance of alien crabs found in the stomach of specimens sampled in the small harbor can be the direct result of the different abundance of *P. gibbesi* in the two areas investigated, and point out the preference of the invasive crab for areas characterized by the presence of breakwaters and/or boulders. Indeed, some authors demonstrated the preference of the invasive crab for harbor's breakwaters, where diversity levels are generally low and where it can find a suitable habitat among breakwaters and boulders [39]. However, from our data, it was not possible to determine whether the goby selectively preys on the crab, or if the abundance of *P. gibbesi* found in the stomachs of *G. paganellus* was related to the greater presence of *P. gibbesi* in the waters of the small harbor.

In all cases, despite the relatively high value of the frequency of occurrence obtained for *P. gibbesi*, the overall diet of *G. paganellus* included a quite wide variety of benthic invertebrates, mainly consisting of small benthic crustaceans. Hence, the Rock Goby may be generally considered as an opportunistic carnivorous species. In other words, the narrow trophic niche indicated by the Levins' index is the result of the abundance of *P. gibbesi* in the stomachs of specimens of *G. paganellus* from the small semi natural harbor, otherwise *G. paganellus* can be considered as a "generalist feeder", as usually reported in literature. After *P. gibbesi*, other crab species, such as *Pisidia* sp., *Xantho* sp., and *P. marmoratus* (juveniles) and the hermit crab *C. erythropus* represent the most important prey types of *G. paganellus*. On the other hand, juveniles of *E. verrucosa* were less represented. Other relatively well represented groups of crustaceans were Amphipoda, Caridea, and Isopoda. Gastropods and polychaetes were rarely found. Finally, algae and seagrasses were probably accidentally ingested, as suggested by some authors for algae [24]. Our results are in general agreement with those reported in literature [12,21,24,26] that describe the diet of *G. paganellus* as mainly composed of small benthic crustaceans, although with some differences among the crustacean groups recorded.

Considering the size range of the analyzed specimens, all the studied individuals of *G. paganellus* were of medium and large size. Indeed, no individual smaller than 7.1 cm TL was recorded. The total length–weight relationship showed an isometric growth, while a positive allometric growth was recorded in specimens from a coastal lagoon of Spain [53]. The analysis of the total length–standard length relationship demonstrated that the SL is on average 83.16% of the TL.

Our study demonstrates for the first time how a native fish species can effectively prey on this invasive crab, reducing the abundance of the established population. Although *P. gibbesi* is quite elusive, *G. paganellus*, a strictly benthic goby of rocky and mixed bottoms, is able to predate on it among rocks and boulders, especially in areas in which the crab is particularly abundant and probably other resources are scarce. Further studies on a large spatio-temporal scale and on the ecology and mechanisms of predation and prey selection are necessary to better understand the role that can have predators such as *G. paganellus* and other native intertidal and shallow waters fish species on the Mediterranean population of *P. gibbesi*. Furthermore, considering the sizes of the preys found and the mouth opening of *G. paganellus*, a medium sized species of goby, the species is able to prey only on juvenile specimens of the crab, while species of larger size and with a wider mouth opening (e.g., *Gobius cobitis*, *Muraena helena*, *Scorpaena maderensis*, *Scorpaena porcus*, *Serranus cabrilla*, *Serranus scriba*) should be investigated for their ability to feed also on adult specimens of *P. gibbesi*. In this regard, it is interesting to note how Sciberras and Schembri [31] reported that in the nearby Maltese waters, the juveniles of *P. gibbesi* were observed between the end of September until at least March, a period that overlaps with that of our study. This would suggest how in Sicily the recruitment could also occur in this period and probably extend for a similar period too. Further studies mentioned above could also clarify this point. To further confirm what was also reported by the same authors, namely the habitat overlap between *P. gibbesi* and *P. marmoratus* in particular, but also with *E. verrucosa*, we also found these two latter species in the stomach content of *G. paganellus*, and *P. marmoratus* was more common than *E. verrucosa*. Another interesting aspect concerning the native predators' control of the population of the invasive *P. gibbesi* was recently pointed out by Noè et al., 2018 [7]. Results obtained by the authors suggested how in areas with high diversity and an abundance of potential predators (e.g., marine protected areas) of *P. gibbesi*, the abundance of the invasive crab is generally lower than in non-protected areas where potential predators are scarcely represented. Other studies are needed in order to clarify the relationships and dynamics between *P. gibbesi* and the presence and abundance of its actual predators.

Invasive species can alter marine ecosystems and cause biodiversity loss, economic damages affecting areas such as fisheries, tourism, and recreational activities, and a threat to human health. Especially in marine environments, totally eradicating or stopping the spread of an invasive alien species have never been successful, and the only valid alternative to date is to limit the spread and abundance through targeted actions [3–5]. For some palatable or even edible alien species, direct actions such as fisheries for human consumption can be a valid solution in order to reduce population. In contrast, when we face non-edible invasive species such as the case of *P. gibbesi*, the promotion and protection of natural predators of invasive species, and the general promotion of the "biotic resistance" (that is, the ability of native species in a community to limit and control the invasion of other species), can be valid additional resources in order to control invasive populations of alien species [7,54,55]. Several studies have already shown how some native species prey on non-indigenous ones, and, in some cases, they are capable of effectively controlling their populations [8,34–36]. Although further studies are necessary in order to better understand this and other predator (native)–prey (IAS) interactions, *G. paganellus* seems to perform the function of "autoregulator" of *P. gibbesi* abundance. Indeed, considering the opportunistic and generalist feeding behavior of *G. paganellus* (and of gobies in general), it preys on the crab especially when the latter is very abundant.

Invasive species can replace or limit the abundance of native ones, alter the habitat structure, and interfere with biogeochemical processes [8,56]. As mentioned above, in a marine environment, the eradication of IAS is complicated, and no eradication attempts have so far been successful. Hence, prevention methods should be prioritized over those of eradication. In this regard, the "biotic resistance" provided by a well-structured native community can act as a natural barrier in preventing or slowing biological invasions. Indeed, on the basis of the mechanisms of competition, IAS cannot establish in a niche similar to that of native species [57]. Furthermore, the occupation of niches is more complete in a species-rich and well-structured native community, and this further contributes to preventing biological invasions [58]. In this regard, the role of native predators, such as in the case here presented of *G. paganellus*, can effectively contribute to controlling populations of invasive species [8]. Although we have no data to state if some specimens of the Rock Goby can learn to prey selectively on the invasive crab, they can certainly contribute to reducing their number. Furthermore, the discovery of the existence of a combination of native predators of the crab would show how, overall and at least in some locations, they could significantly contribute to controlling and limiting the population's growth of the invasive crab. Particular measures could also be promoted in order to protect the population of key native species from human activities.

5. Conclusions

In conclusion, our study demonstrates how the native fish *G. paganellus* can prey on the invasive crab *P. gibbesi*, being able to contribute to the control of the population of this IAS in the Mediterranean Sea. Although *P. gibbesi* was the dominant prey type, the diet of *G. paganellus* was composed of a wide variety of invertebrates, mainly represented by small benthic crustaceans, brachyurans in particular. Hence, we can consider this species as an opportunistic feeder able to prey on *P. gibbesi* too, thus potentially playing a role in the biological control of the species. However, more studies are needed to understand if the native goby can (and under what conditions) develop the ability to selectively feed on *P. gibbesi*. Furthermore, other studies can help to expand this research to other species and areas in order to evaluate the overall impact of predation of the coastal fish community on this invasive crab (and other invasive species), and relate the presence of the major predators of *P. gibbesi* to its abundance levels at several locations. This study also provided the first biological and ecological data on *G. paganellus* from specimens from the Ionian Sea (central Mediterranean Sea).

The lack of predators and/or parasites can cause overpopulation of invasive species in their new colonized areas, especially if these species cannot be exploited economically by fishery because they have no economic value. *Percnon gibbesi* falls in this latter category and showed a great success in colonizing the Mediterranean Sea, with potentially negative although unpredictable effects on the ecosystem.

To date, no studies had shown a direct predation on this crab by a native species. Although our study clearly demonstrated the predation of the native *G. paganellus* on the invasive crab *P. gibbesi*, further studies are still needed in order to better understand the mechanisms of this predator–prey interaction, the degree of contribution that this predation can have in controlling the abundance of this invasive species, and if other Mediterranean populations of the goby have adapted or are able to prey on this invasive crab. Furthermore, it is very likely that other species as well, such as some coastal fish, prey on *P. gibbesi*, and other studies should focus on finding out which ones and to what extent they can contribute overall to the control of the invasive Mediterranean population of the crab.

Author Contributions: Conceptualization: F.T.; methodology: F.T., G.M. and B.M.L.; investigation: F.T. and G.M., formal analysis: F.T., G.M. and B.M.L.; writing—original draft preparation: F.T. and G.M.; writing—review and editing: F.T., G.M. and B.M.L. All authors have read and agreed to the published version of the manuscript.

Funding: This research received no external funding.

Institutional Review Board Statement: Ethical review and approval are not necessary for this study because collected specimens come from fishing activities and have been treated ethically.

Informed Consent Statement: Not applicable.

Data Availability Statement: Data will be made available on request.

Acknowledgments: Authors are grateful to Giacomo Bernardi, Department of Ecology and Evolutionary Biology, University of California Santa Cruz, for the English language review, and to the three anonymous reviewers for their helpful comments.

Conflicts of Interest: The authors declare no conflict of interest.

References

1. Ghosholz, E. Ecological and evolutionary consequences of coastal invasions. *Trends Ecol. Evol.* **2002**, *17*, 22–27. [CrossRef]
2. Molnar, J.L.; Gamboa, R.L.; Revenga, C.; Spalding, M.D. Assessing the global threat of invasive species to marine biodiversity. *Front. Ecol. Environ.* **2008**, *6*, 485–492. [CrossRef]
3. Bellard, C.; Cassey, P.; Blackburn, T.M. Alien species as a driver of recent extinctions. *Biol. Lett.* **2016**, *12*, 20150623. [CrossRef]
4. Ricciardi, A.; Palmer, M.E.; Yan, N.D. Should biological invasions be managed as natural disasters? *BioScience* **2011**, *61*, 312–317. [CrossRef]

5. Bailey, S.; Brown, L.; Campbell, M.; Canning-Clode, J.; Carlton, J.; Castro, N.; Chainho, P.; Chan, F.T.; Creed, J.C.; Curd, A.; et al. Trends in the detection of aquatic non-indigenous species across global marine, estuarine and freshwater ecosystems: A 50-year perspective. *Divers. Distrib.* **2020**, *26*, 1780–1797. [CrossRef]
6. Hooper, D.U.; Chapin, F.S.; Ewel, J.J.; Hector, A.; Inchausti, P.; Lavorel, S.; Lawton, J.H.; Lodge, D.M.; Loreau, M.; Naeem, S.; et al. Effects of biodiversity on ecosystem functioning: A consensus of current knowledge. *Ecol. Monogr.* **2005**, *75*, 3–35. [CrossRef]
7. Noè, S.; Gianguzza, P.; di Trapani, F.; Badalamenti, F.; Vizzini, S.; Fernández, T.V.; Bonaviri, C. Native predators control the population of an invasive crab in no-take marine protected areas. *Aquatic Conserv. Mar. Freshw. Ecosyst.* **2018**, *28*, 1229–1237. [CrossRef]
8. Madenjian, C.P.; Stapanian, M.A.; Witzel, L.D.; Einhouse, D.W.; Pothoven, S.A.; Whitford, H.L. Evidence for predatory control of the invasive round goby. *Biol. Invasions* **2011**, *13*, 987–1002. [CrossRef]
9. Fricke, R.; Eschmeyer, W.N.; Fong, J.D. Eschmeyer's Catalog of Fishes: Species by Family/Subfamily. California Academy of Sciences, San Francisco, USA. Available online: http://researcharchive.calacademy.org/research/ichthyology/catalog/SpeciesByFamily.asp (accessed on 17 November 2020).
10. Nelson, J.S.; Grande, T.C.; Wilson, M.V.H. *Fishes of the World*, 5th ed.; John Wiley and Sons: Hoboken, NJ, USA, 2016; p. 752.
11. Kovačić, M. Checklist of gobies (Teleostei: Gobiidae) of the Mediterranean Sea and a key for species identification. *Zootaxa* **2020**, *4877*, 75–101. [CrossRef]
12. Hajji, F.; Ouannes-Ghorbel, A.; Ghorbel, M.; Jarboui, O. Diet shifts of *Gobius paganellus* (Teleostei, Gobiidae) from the Gulf of Gabes (Central Mediterranean). *Vie Milieu* **2012**, *62*, 69–76.
13. Hajji, F.; Ouannes-Ghorbel, A.; Ghorbel, M.; Jarboui, O. Reproductive biology of the rock goby, *Gobius paganellus* (Actinopterygii: Perciformes: Gobiidae), on the southern Tunisian coast (Gulf of Gabes). *Cienc. Mar.* **2012**, *38*, 505–515. [CrossRef]
14. Kovačić, M.; la Mesa, A. Feeding ecology of the Buen's goby, *Buenia affinis*, in the Kvarner area (Adriatic Sea). *Vie Milieu* **2008**, *58*, 277–281.
15. Tiralongo, F.; Messina, G.; Lombardo, B.M. First data on habitat preference, diet and length-weight relationship of *Gobius incognitus* Kovačić & Šanda, 2016 (Pisces: Gobiidae). *Acta Adriat.* **2020**, *61*, 67–78. [CrossRef]
16. Iglésias, S.P.; Vukić, J.; Sellos, D.Y.; Soukupová, T.; Šanda, R. *Gobius xoriguer*, a new offshore Mediterranean goby (Gobiidae), and phylogenetic relationships within the genus *Gobius*. *Ichthyol. Res.* **2021**, 1–15. [CrossRef]
17. Kovačić, M.; Šanda, R. A new species of *Gobius* (Perciformes: Gobiidae) from the Mediterranean Sea and the redescription of *Gobius bucchichi*. *J. Fish Biol.* **2016**, *88*, 1104–1124. [CrossRef]
18. Kovačić, M.; Patzner, R.; Schliewen, U. A first quantitative assessment of the ecology of cryptobenthic fishes in the Mediterranean Sea. *Mar. Biol.* **2012**, *159*, 2731–2742. [CrossRef]
19. Miller, P.J. Gobiidae. In *Fishes of the North-Eastern Atlantic and Mediterranean*; Whitehead, P.J.P., Bauchot, M.L., Hureau, J.C., Nielsen, J., Tortonese, E., Eds.; UNESCO: Paris, France, 1986; pp. 1019–1085.
20. Tortonese, E. Mediterranean fishes present in the Red Sea: Pan-oceanic and anti-lessepsian species. *Cybium* **1984**, *8*, 99–102.
21. Engin, S.; Seyhan, K. Biological Characteristics of rock goby, *Gobius paganellus* (Actinopterygii: Perciformes: Gobiidae), in the South-Eastern Black Sea. *Acta Ichthyol. Pisc.* **2009**, *39*, 111–118. [CrossRef]
22. Miller, P.J. Age, growth, and reproduction of the rock goby, *Gobius paganellus* L., in the Isle of Man. *J. Mar. Biol. Assoc. UK* **1961**, *41*, 737–769. [CrossRef]
23. Wilkins, H.K.A.; Myers, A.A. Microhabitat utilisation by an assemblage of temperate Gobiidae (Pisces: Teleostei). *Mar. Ecol. Prog. Ser.* **1992**, *90*, 103–112. [CrossRef]
24. Azevedo, J.M.N.; Simas, A.M.V. Age and growth, reproduction and diet of a sublittoral population of the rock goby *Gobius paganellus* (Teleostei, Gobiidae). *Hydrobiologia* **2000**, *440*, 129–135. [CrossRef]
25. Louiz, I.; Attia, M.B.; Hassine, O.K.B. Some aspects of reproductive biology of *Gobius paganellus* (Gobiidae) on the north-eastern coasts of Tunisia (Bizerta lagoon). *J. Mar. Biol. Assoc. UK* **2013**, *93*, 2235–2246. [CrossRef]
26. Mazé, R.A. Seasonal and ontogenetic diet shift in an intertidal population of *Gobius paganellus* (Teleostei, Gobiidae) from the Cantabrian Coast. *Vie Milieu* **2004**, *54*, 1–6.
27. Katsanevakis, S.; Poursanidis, D.; Yokes, M.B.; Mačić, V.; Beqiraj, S.; Kashta, L.; Sghaier, Y.R.; Zakhama-Sraieb, R.; Benamer, I.; Bitar, G.; et al. Twelve years after the first report of the crab *Percnon gibbesi* (H. Milne Edwards, 1853) in the Mediterranean: Current distribution and invasion rates. *J. Biol. Res.* **2011**, *16*, 224–236.
28. Relini, M.; Orsi, L.; Puccio, V.; Azzurro, E. The exotic crab *Percnon gibbesi* (H. Milne Edwards, 1853) (Decapoda, Grapsidae) in the Central Mediterranean. *Sci. Mar.* **2000**, *64*, 337–340. [CrossRef]
29. Suaria, G.; Pierucci, A.; Zanello, P.; Fanelli, E.; Chiesa, S.; Azzurro, E. *Percnon gibbesi* (H. Milne Edwards, 1853) and *Callinectes sapidus* (Rathbun, 1896) in the Ligurian Sea: Two additional invasive species detection made in collaboration with local fishermen. *BioInvasions Rec.* **2017**, *6*, 147–151. [CrossRef]
30. Pipitone, C.; Badalamenti, F.; Sparrow, A. Contribution to the knowledge of *Percnon gibbesi* (Decapoda, Grapsidae), an exotic species spreading rapidly in Sicilian waters. *Crustaceana* **2001**, *74*, 1009–1017.
31. Sciberras, M.; Schembri, P.J. Biology and interspecific interactions of the alien crab *Percnon gibbesi* in the Maltese Islands. *Mar. Biol. Res.* **2008**, *4*, 321–332. [CrossRef]

32. Crocetta, F.; Shokouros-Oskarsson, M.; Doumpas, N.; Giovos, I.; Kalogirou, S.; Langeneck, J.; Tanduo, V.; Tiralongo, F.; Virgili, R.; Kleitou, P. Protect the natives to combat the aliens: Could *Octopus vulgaris* Cuvier, 1797 be a natural agent for the control of the lionfish invasion in the Mediterranean Sea? *J. Mar. Sci. Eng.* **2021**, *9*, 308. [CrossRef]
33. Kaldonski, N.; Lagrue, C.; Motreuil, S.; Rigaud, T.; Bollache, L. Habitat segregation mediates predation by the benthic fish *Cottus gobio* on the exotic amphipod species *Gammarus roeseli*. *Naturwissenschaften* **2008**, *95*, 839–844. [CrossRef]
34. Maljković, A.; van Leeuwen, T.E.; Cove, S.N. Predation on the invasive red lionfish, *Pterois volitans* (Pisces: Scorpaenidae), by native groupers in the Bahamas. *Coral Reefs* **2008**, *27*, 501. [CrossRef]
35. Louette, G. Use of a native predator for the control of an invasive amphibian. *Wildl. Res.* **2012**, *39*, 271–278. [CrossRef]
36. Giakoumi, S.; Pey, A.; Thiriet, P.; Francour, P.; Guidetti, P. Patterns of predation on native and invasive alien fish in Mediterranean protected areas. *Mar. Environ. Res.* **2019**, *150*, 104792. [CrossRef]
37. Yokes, B.; Galil, B.S. Touchdown—First record of *Percnon gibbesi* (H. Milne Edwards, 1853) (Crustacea: Decapoda: Grapsidae) from the Levantine coast. *Aquat. Invasions* **2006**, *1*, 130–132. [CrossRef]
38. Galil, B.; Froglia, C.; Noel, P.Y. Atlas of Exotic Species in the Mediterranean. 2008. Available online: www.ciesm.org/atlas/appendix2.html (accessed on 10 December 2020).
39. Deudero, S.; Frau, A.; Cerda, M.; Hampel, H. Distribution and densities of the decapod crab *Percnon gibbesi*, an invasive Grapsidae, in western Mediterranean waters. *Mar. Ecol. Prog. Ser.* **2005**, *285*, 151–156. [CrossRef]
40. Cannicci, S.; Badalamenti, F.; Milazzo, M.; Gomei, M.; Baccarella, A.; Vannini, M. Unveiling the secrets of a successful invader: Preliminary data on the biology and the ecology of the crab *Percnon gibbesi* (H. Milne Edwards, 1853). *Rapp. Com. Int. Mer Médit.* **2004**, *37*, 326.
41. Jones, R.E.; Petrell, R.J.; Pauly, D. Using modified length-weight relationships to assess the condition of fish. *Aquac. Eng.* **1999**, *20*, 261–276. [CrossRef]
42. Froese, R. Cube law, condition factor and weight-length relationships: History, meta-analysis and recommendations. *J. Appl. Ichthyol.* **2006**, *22*, 241–253. [CrossRef]
43. Riedl, R. *Fauna e Flora del Mediterraneo*; Franco Muzzio Editore: Padua, Italy, 2010; p. 777.
44. Hayward, P.J.; Ryland, J.S. *Handbook of the Marine Fauna of North-West Europe*; Oxford University Press: Oxford, UK, 1990; p. 816.
45. Hyslop, E.J. Stomach content analysis, a review of methods and their application. *J. Fish Biol.* **1980**, *17*, 411–429. [CrossRef]
46. Carrassòn, M.; Matallanas, J.; Casadevall, M. Feeding strategies of deep-water morids on the western Mediterranean slope. *Deep Sea Res.* **1997**, *44*, 1685–1699. [CrossRef]
47. N'da, K. Regime alimentaire du rouget de roche *Mullus surmuletus* (Mullidae) dans le nord du golfe de Gascogne. *Cybium* **1992**, *16*, 159–168.
48. Krebs, J.C. *Ecological Methodology*; Harper & Row: New York, NY, USA, 1989; p. 620.
49. Novakowski, G.C.; Hahn, N.S.; Fugi, R. Diet seasonality and food overlap of the fish assemblage in a pantanal pond. *Neotrop. Ichthyol.* **2008**, *6*, 567–576. [CrossRef]
50. Brown, S.C.; Bizzarro, J.J.; Cailliet, G.M.; Ebert, D.A. Breaking with tradition: Redefining measures for diet description with a case study of the Aleutian skate *Bathyraja aleutica* (Gilbert 1869). *Environ. Biol. Fish.* **2012**, *95*, 3–20. [CrossRef]
51. R Core Team. R: A Language and Environment for Statistical Computing. 2018. Available online: http://www.R-project.org (accessed on 10 December 2020).
52. Zander, C.D.; Berg, J. Feeding ecology of littoral gobiid and blennioid fishes of the Banyuls area (Mediterranean Sea) II. Prey selection and size preference. *Vie Milieu* **1984**, *34*, 149–157.
53. Cubedo-Verdiell, D.; Oliva-Paterna, F.J.; Torralva, M. Length-weight relationships for 22 fish species of the Mar Menor coastal lagoon (western Mediterranean Sea). *J. Appl. Ichthyol.* **2006**, *22*, 293–294. [CrossRef]
54. Giakoumi, S.; Katsanevakis, S.; Albano, P.G.; Azzurro, E.; Cardoso, A.C.; Cebrian, E.; Deidun, A.; Edelist, D.; Francour, P.; Jimenez, C.; et al. Management priorities for marine invasive species. *Sci. Total Environ.* **2019**, *688*, 976–982. [CrossRef]
55. Byun, C.; Lee, E.J. Ecological application of biotic resistance to control the invasion of an invasive plant, *Ageratina altissima*. *Ecol. Evol.* **2017**, *7*, 2181–2192. [CrossRef] [PubMed]
56. Mack, R.N.; Simberloff, D.; Lonsdale, W.M.; Evans, H.; Clout, M.; Bazzaz, F.A. Biotic invasions: Causes, epidemiology, global consequences, and control. *Ecol. Appl.* **2000**, *10*, 689–710. [CrossRef]
57. Funk, J.L.; Cleland, E.E.; Suding, K.N.; Zavaleta, E.S. Restoration trough reassembly: Plant traits and invasion resistance. *Trends Ecol. Evol.* **2008**, *23*, 695–703. [CrossRef]
58. Elton, C.S. *The Ecology of Invasions by Animals and Plants*; The University of Chicago Press: Chicago, IL, USA, 1958.

Journal of
Marine Science and Engineering

MDPI

Article

Relative Influence of Environmental Factors on Biodiversity and Behavioural Traits of a Rare Mesopelagic Fish, *Trachipterus trachypterus* (Gmelin, 1789), in a Continental Shelf Front of the Mediterranean Sea

Armando Macali [1,*], Alexander Semenov [2], Francesco Paladini de Mendoza [3], Alessia Dinoi [4], Elisa Bergami [5] and Francesco Tiralongo [3,6]

1 Department of Ecological and Biological Sciences, Ichthyogenic Experimental Marine Center (CISMAR), University of Tuscia, 01016 Tarquinia, Italy

2 N. A. Pertsov White Sea Biological Research Station, Department of Biology, Lomonosov Moscow State University, 119991 Moscow, Russia; semenov@wsbs-msu.ru

3 Ente Fauna Marina Mediterranea, 96012 Avola, Italy; framendoza1985@gmail.com (F.P.d.M.); francesco.tiralongo@unict.it (F.T.)

4 Center for Ecological Genomics and Wildlife Conservation, Department of Zoology, University of Johannesburg, Auckland Park 2006, South Africa; alexiadinoi@gmail.com

5 Department of Physical, Earth and Environmental Sciences, University of Siena, Via Mattioli 4, 53100 Siena, Italy; elisa.bergami@unisi.it

6 Department of Biological, Geological and Environmental Sciences, University of Catania, 95129 Catania, Italy

* Correspondence: a.macali@unitus.it

Received: 7 July 2020; Accepted: 30 July 2020; Published: 2 August 2020

check for updates

Abstract: Coastal environments can be influenced by water body masses with particular physical, chemical, and biological properties that create favourable conditions for the development of unique planktonic communities. In this study, we investigated a continental shelf front at Ponza Island (Tyrrhenian Sea) and discussed its diversity and complexity in relation to major environmental parameters. Moon phase and current direction were found to play a significant role in shaping species abundance and behaviour. During in situ observations, we also provided the first data on the behaviour of juveniles of a rare mesopelagic species, *Trachipterus trachypterus*, suggesting the occurrence of Batesian mimicry.

Keywords: Trachipteridae; Ponza Island; upwelling; plankton diversity; Batesian mimicry

1. Introduction

Open water environments are highly dynamic ecosystems [1], with ecological processes spreading over a wide range of spatial and temporal scales [2]. At a meso- and sub-mesoscale (~1–100 km), these processes emerge as dominant structuring regimes for populations of marine organisms in the short time span (days to months) [3,4], encompassing key biological dynamics, such as bloom lifetime or behavioural switches of marine predators [5]. These ecological processes are particularly relevant for drifting organisms (plankton) that are advected and dispersed by the ocean currents [6]. Within this complex scenario, sampling probabilities for uncommon and little-known species can be higher than those reported in all other marine environments, in which these species are typically rare and occur at low abundances. This may result in uncertainty about the real species abundances and their ecological traits. As a result, modern biodiversity surveillance programs often investigate the occurrence of species mostly present at moderate-to-high abundance level [7,8].

The complex interactions between coastal morphology and sea surface circulation of the Mediterranean Sea promote a high degree of ecosystem patchiness [9]. The environmental characteristics (e.g., salinity and temperature), as well as the dynamics of enrichment processes in the water column [5], lead to differences in distribution and abundance of pelagic species [10–12]. While marine macro-ecology has benefited from the analysis of spatially extensive data sites, inferences about ecological processes are better evaluated by dynamic data collection over temporal scales on a smaller geographic range [13]. Animal behaviour also affect assemblage composition, as for species displaying avoidance behaviour (e.g., due to increased predation risk) or with small home ranges, which are less likely to be observed than schooling or highly mobile species [14].

Within the Mediterranean fish biodiversity, little is still known on mesopelagic communities. Among deep water species, representatives of the family Trachipteridae Swainson, 1839 (order Lampriformes) are rare encounters. They are represented by 10 species distributed in 3 genera [15]: *Desmodema* (Walters and Fitch, 1960), *Trachipterus* (Goüan, 1770), and *Zu* (Walters and Fitch, 1960) [16]. Of these, only *Trachipterus arcticus* (Brünnich, 1788), *Trachipterus trachypterus* (Gmelin, 1789) and *Zu cristatus* (Bonelli, 1819) have been reported in the Mediterranean Sea. The former is a North Atlantic species, with a single record in the Mediterranean Sea from Spain [17]. The other two species, *T. trachypterus* and *Z. cristatus*, reported from tropical and temperate waters of all oceans, have been found in several Mediterranean areas [18–27]. In the Mediterranean Sea, fish of this family are accidentally caught with professional fishing gears, but always in low numbers [20,25,28]. Given the complexity of mesopelagic environments, an understanding of the triggers for the distribution of rare mesopelagic species requires a more comprehensive knowledge of the temporal pattern of environmental variation in a variety of timescales. For this reason, long-term environmental monitoring at aggregation sites is of critical importance.

With the present study, we examined the extent to which key features of the hydrodynamic environment predict the presence of *T. trachypterus* in a Mediterranean continental shelf front over a 21-days timescale. Using multivariate hierarchical analysis, we examined the temporal changes in species occurrence at an aggregation site. Field observations over time have allowed us to discuss on the drivers affecting the occurrence at particular times and on some relevant ecological and behavioural traits of these species.

2. Materials and Methods

2.1. Study Area and Sampling Design

The study was performed in the framework of the AQUATILIS EXPEDITION at Ponza Island (Tyrrhenian Sea, Italy) from the 8th to the 28th of April 2018. Due to its bathymetric profile, the area is exposed to prominent surface and upwelling filaments, resulting in a complex hydrodynamism that favours the accumulation of planktonic species in patched areas, according to the prevalent weather forcing [29]. Diving sites were close to "Secca delle Formiche" (40.884947° N, 12.978043° E), a group of emerged rocks, located 0.8 nautical miles (nm) SE from the port of Ponza. The bank rapidly sinks from a depth of 5 m to over 500 m (Figure 1), representing a suitable place to study the upwelling of deep mesopelagic species.

Previous surveys conducted in April 2016 and 2017 highlighted a remarkable difference in plankton abundance and biodiversity in the study area according to time; in particular, observation during night dives showed a more complex plankton biodiversity, if compared with day dives (authors' personal note). As a consequence, in the present study, scuba night dives have been planned to start one hour after sunset. Two different groups of three divers each equipped with Nikon D850 with 50 mm f2 Zeiss Milvus macro lens for photography, Panasonic Lumix GH5 with Panasonic Lumix G 8 mm f3.5 fish eye and Panasonic Lumix 15 mm f1.7 lenses for video and tools (fine mesh hand nets, 250 mL plastic cans) for sampling collection, performed alternate dives. In order to increase the image quality, all the cameras were equipped with a buoyancy stabilization system. Sampling depth varied between 0.5 and 10 m, freely drifting within the surface current. To ensure taxon identity, 4 specimens of *T. trachypterus* were collected and analyzed for morphological traits. Behaviour was recorded by filming all the specimens detected. Observations followed the same schedule, starting one hour after sunset, freely drifting in the first 10 m of the water column. Surveys were conducted daily between 21:00 and 24:00 h.

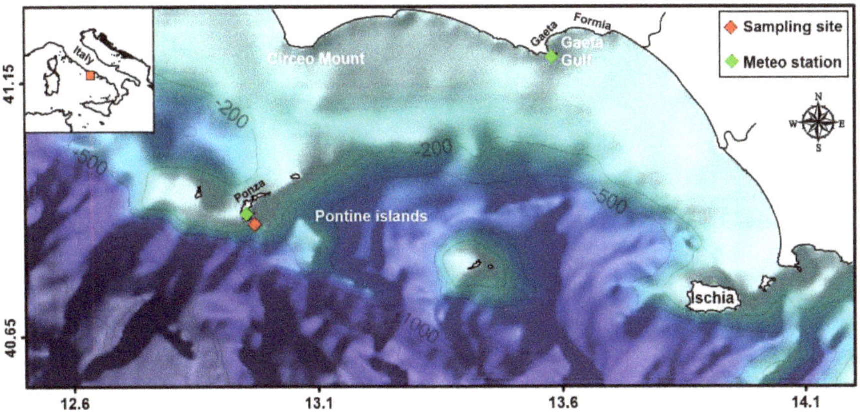

Figure 1. Regional view of the coastal site with the localization of the study area (Ponza Island, within the Pontine Island archipelago) and weather station used for the physical process analysis.

2.2. Environmental Data Collection

Main abiotic environmental conditions were inferred from numerical models and in situ measurements (current direction). To this aim, wind records from available meteorological station and numerical results of Surface Sea Temperature (SST) and Salinity (SS) were collected considering both surveying period and long-time series. The wind records were downloaded from the Gaeta Port Authority and the Mareographic network websites (www.mareografico.it; www.portidiroma.it). Ponza's records extend only for a 4-year period (2011–2015), while more than 10-year records are available from Gaeta (2007–2018) (Supplementary Materials, Figure S1). Despite the distance between stations, wind roses can be considered comparable [29,30]. During surveys, we used Gaeta Meteo Station period to analyse wind conditions. The SST and SS datasets were downloaded from the "Copernicus" data portal (http://marine.copernicus.eu, accessed on the 21st of November 2018). SST and SS from 1987 to present were analysed, producing long-term temporal mean distribution of variables during April months and mean conditions during the surveyed period.

Moonlight is known to affect zooplankton and micro-nekton dynamics, as well as their predators in tropical, subtropical, and Arctic waters [31–36]. Moonlight also drives vertical migrations to bathypelagic depths (~1000 m) involving a cascade response, with deeper-living organisms responding to movements of shallower-living, surface-synchronized populations [37]. According to this evidence, the main macro-zooplanktonic species composition was recorded and correlated to moon phases and moonlight intensity.

2.3. Data Analysis

Non-parametric multivariate analyses were performed to assess the relation between the species observed and the environmental parameters over the survey time period. To analyze the environmental conditions (SST, moon phase, wind direction, and current direction), the parameters were normalized and standardized to carry out a Principal Component Analysis (PCA) with Statistical Package for Social Science (SPSS, Version 26.0, IBM Corp© 2019). Once the significant environmental patterns were established, standardized data of the observed species were used to investigate the variability of these patterns. To assess variations in the species diversity, data of the thirty-five taxa were analyzed using distance-based permutational multivariate analysis of variance (PERMANOVA) [38,39] with the method of permutation of the residuals under a reduced model, according to a 2-factor nested design: moon phase and current direction. The homogeneity of multivariate dispersions was tested with the PERMDISP routine, mainly to estimate dispersion at the moon phase level. Non-parametric multidimensional scaling ordination (nMDS) was used to examine the behaviour of the species grouped according to the current direction pattern. The Similarity Percentage analysis procedure (SIMPER) was used to identify the main species that characterized each current direction pattern. These analyses were performed on Primer v.7 with PERMANOVA+ software [39,40]. A PCA with supplementary variables was performed with CANOCO 5 [41] in order to investigate the species response to each variable. Finally, the hierarchical cluster analysis was applied to estimate the number of clusters using Primer v.7 [40].

3. Results

3.1. Oceanographic Landscape

Available time-lapse records of Ponza Island and Gaeta showed three main directions (North-East, South-East and West-Northwest) of wind. Between the 12th and the 18th of April 2018, a moderate north-west wind dominated, while in the other days, weak northern flows (8th–11th of April) and eastern flows (19th–28th of April) were recorded (Supplementary Materials, Figure S2B). The SST mean distribution of April showed a quite uniform temperature at a regional scale, with variations of less than 1 °C (16.8–17.6 °C). Between the 9th and the 18th of April 2018, the SST was generally cooler than the mean long-term map, with a weak NW-SE increasing gradient. In correspondence of the study area, the SST reached 15.3–15.5 °C. After the 18th of April, a sensitive increase (more than 2 °C) in SST occurred at a regional scale (Supplementary Materials, Figure S2A). SS mean distribution map (Supplementary Materials, Figure S2C) shows the effect of continental waters (Volturno River) which reduced SS in the nearshore. In the study site, the SS reached 38 psu on the 9th–11th of April 2018, slowly decreasing until the end of the survey (Supplementary Materials, Figure S2C).

3.2. Assemblages

A total of 35 families of meso- and macroplankton taxa has been recorded over the whole studied period, together with the associated environmental factors (Supplementary Materials, Table S1). The species list included pelagic predators (*Forskaliidae, Ommastrephidae, Pelagidae,* and *Phronimidae*) as well as planktivorous species (*Alciopydae, Fritillariidae, Leucotheidae, Mysidae,* and *Salpidae*) representing exploiting food characteristics of the current fronts. Details of the trophic and reproductive behaviour of these species are reported in the Supplementary Materials (Plate I-III).

3.3. Role of Environmental Factors

The PCA model computed with SPSS explained more than 90% of the variability for the considered environmental variables (Figure 2). Based on these results and the PERMANOVA, two environmental patterns were retained as main factors: the moon phase (p-value = 0.0011) and the current direction (p-value = 0.0015). The SST did not play a significant role in species composition and diversity, while showing high variability (range 14.5–19 °C) in the surveyed period.

Moonlight seems to rule the occurrence and abundance of plankton in the surface layer of the water column, with a clear increase in the occurrence of 12 taxa, consequently to the extremes of the moon phase (see Supplementary Materials, Figure S3). With the exception of Cestidae, all the other taxa included in this group showed a negative correlation to moonlight, as for Nereidae and Sepiolidae, which was also correlated to the record of mating behaviour (see Supplementary Materials, Plate I-II). Juvenile and larval stages of benthic and benthopelagic species, as Pleuronectiformes, Phycidae, Anguilliformes, and Scyllaridae, also negatively responded to moonlight.

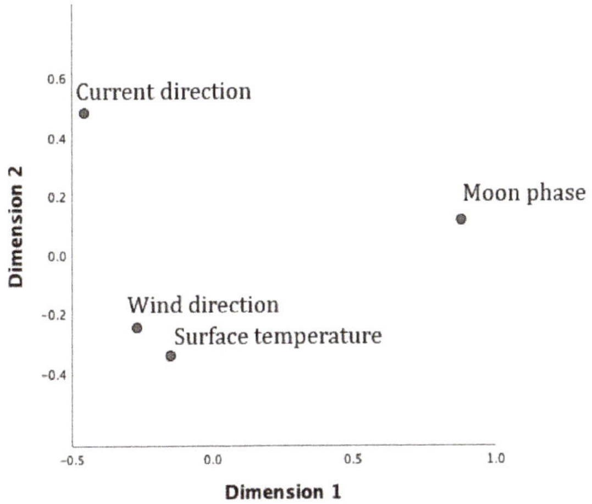

Figure 2. Principal component analysis (PCA) of the environmental factors (performed with SPSS).

According to the observed pattern, the most abundant taxon during the main Eastern flow direction was represented by Forskaliidae, while in the least North-Western flow the most abundant taxon was represented by Alciopidae.

The PCA model performed with CANOCO 5 shows correlations among moon phase, current directions, and taxa. It is possible to observe how most taxa took the opposite direction as the moon phase progressed and how the main current blows eastward (Figure 3a). Arrows indicate the direction of the taxa at increasing values of the environmental variables, with longer vectors representing a greater range of variation in the observed values (Figure 3b). A stronger correlation is associated to more acute angles between vectors and the axes. The x axis explains 57.6% of the variation among parameters at each taxon, whereas the y axis explains 47.1% (Figure 3a,b).

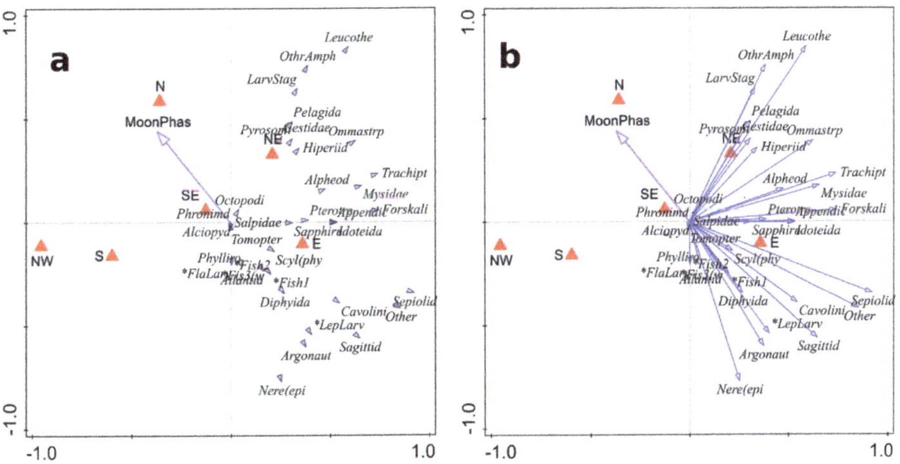

Figure 3. PCA (**a**) of the encountered taxa and the two main environmental factors of moon phase and current directions. PCA with arrows (**b**) highlights the direction of the taxa at increasing values of the environmental variables (performed with CANOCO 5).

The hierarchical cluster analysis performed with Primer v7 shows 5 main clusters based on the taxa presence and the current direction as environmental factor (Figure 4).

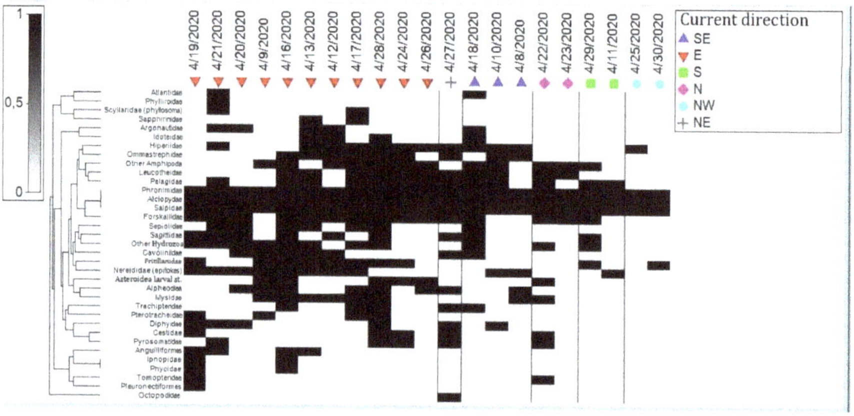

Figure 4. Dendogram plot performed by Primer v7 combined with the shadow plot for the presence (from 0 to 1) in the monitored time and the daily current direction.

According to the statistical analyses on the relation between environmental constrains and species diversity, the contribution of the eastern currents in shaping plankton communities is well supported (see Figure 4). Lowest level of plankton diversity is related to north western stream, which can be considered local, as the diving site results covered by the island from this direction (0.8 nm off the bank). Species diversity mainly reflects typical pelagic assemblages, whereas the occurrence of emergent zooplankton (Amphipoda, Nereidae) and more benthic species (Octopodidae) can be considered mainly influenced by the contribution of the nearby shallow water banks ("Secca delle Formiche") and affected by moon phase rather than current direction. The occurrence of other larval and juvenile stages, as for Anguilliformes, Phycidae, Pleuronectiformes, and Scyllaridae (Figure 4), showed a clear convergence in the contribution of current direction (Figure 4) and moon phase (Supplementary Materials, Figure S3), which may be related to a shared behavioural form of detection avoidance.

3.4. Behavioral Traits of *T. trachypterus*

Over the whole study period, a total of 18 juveniles of *T. trachypterus* were observed, photographed and filmed during SCUBA dives at night between the 16th and the 28th of April 2018. Prolonged observations (>2 min) of the specimens were possible only for 10 fish, given the high avoidance behaviour of the species. Juveniles were observed swimming between 0.5 and 7 m below the surface, with the body obliquely or almost vertically pointed with the head towards the surface. Locomotion was performed through undulating movements of the long dorsal fin only, except for the anterior prolonged dorsal fin rays that showed limited oscillating movements (Supplementary Materials, Video 1–10). A faster escaping movement was displayed using the whole body, as showed in Supplementary Materials, Video 8. Pelvics and caudal fin remained spread, but substantially motionless or with small and occasional "snap" movements. While dorsal and pectoral fin membranes were entirely translucent, those of the wide pelvic and caudal fins showed orange blotches, with rays protruding far over the fin membrane. In some cases, juveniles oriented the dorsal or abdominal surface towards the observer, greatly limiting the area exposed to the view of the observer itself (Figure 5C,D; Supplementary Materials, Video 5–7, 9–10). At the same time, caudal and pelvic fins were maintained orthogonal to the observer, with a significant torsion of the caudal fin up to 90° angle (Figure 5C). This latter behaviour was displayed for a prolonged time.

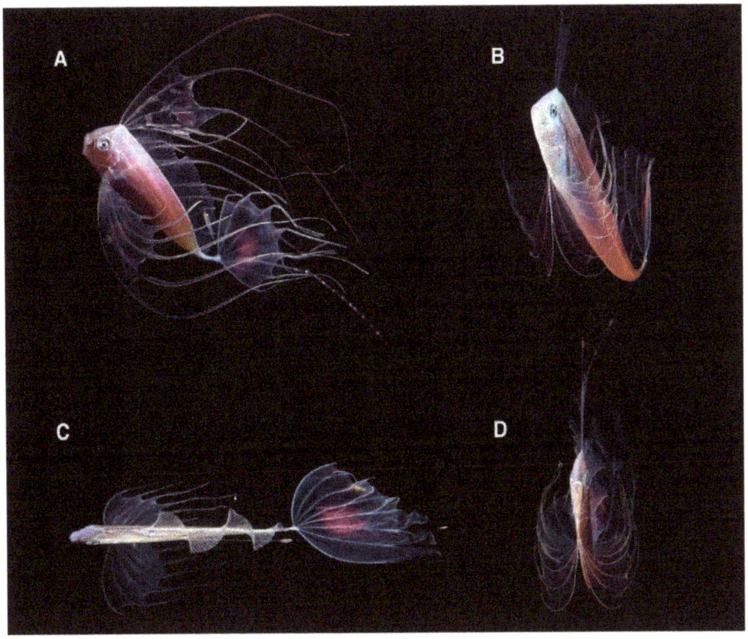

Figure 5. Lateral (**A**), fronto-lateral (**B**), dorsal (**C**), and frontal (**D**) view of juvenile specimens of *T. trachypterus*, with detail of fins orientation accordingly to the observer visual axis.

Fish were also observed rapidly protruding their jaws, probably preying on small planktonic organisms (Supplementary Materials, Video 4), that were very abundant in the surface water column in which specimens were observed.

Four specimens were caught by hand net and preserved in alcohol for laboratory analysis. All the main morphometric parameters are reported in Supplementary Materials, Table S2.

4. Discussion

4.1. Ponza Island Coastal front and Pelagic Biodiversity

Biological–physical interactions structure the variability of the marine environment at a wide range of spatial and temporal scales, affecting population dynamics and trophic interactions [42,43]. Such example is a front, a physical interface with gradients of water properties that include temperature, salinity, and turbidity [44,45]. Fronts occur across the world oceans, ranging from basin-scale features to small river plumes, and can be persistent or ephemeral [44,46]. Frontal zones are often associated with enhanced biomass, and may serve as an important foraging grounds by aggregating species from multiple trophic levels [47–49], being key habitats for successful energy transfer through food webs. In situ observations and modelling studies suggest that mesoscale and sub-mesoscale processes may affect biodiversity in the Mediterranean Sea, especially where coastal morphology and intense wind stress lead to the formation of energetic filaments [50,51], contributing to the dispersal of coastal inputs towards the open sea, along with plankton.

Ponza Island is located in the western sector of the Pontine volcanic archipelago and originated during the Plio-Pleistocene on the outer margin of the Latium continental shelf. The island slope rapidly sinks to deep waters bottoms, especially on the east and south-west side (Figure 1). Winds are known to support the organic enrichment process of offshore waters, bringing the contribution of coastal freshwater input. Available time lapse records of Ponza Island and Gaeta showed three prevalent wind directions, the eastern one of which can be considered the most relevant, both in terms of intensity and frequency. This is clearly shown by the SS map presented in Supplementary Materials, Figures S1C and S2. The overlap of deep water pools, eutrophic waters, and wind driven currents may lead to an organic enrichment in the study site, due to both the upwelling and the contribution of coastal freshwater waters, supporting an increase in plankton biodiversity. Vertical transport has a well known effect in increasing the biodiversity of the photic layers only in restricted areas, as, for example, where convection is sufficiently deep, in a small number of frontal regions and in the few upwelling sites. Coastal inputs have thus an important role in sustaining food webs in the whole Mediterranean Sea [52].

Despite the data acquired in this expedition they cannot be considered exhaustive to completely depict these processes (both in terms of temporal and sampling extension), they show intriguing insights into the relation between local hydrodynamism and plankton diversity. This latter conclusion is enhanced by the record of a number of rare species, some of which appear to be far less rare than thought. An example is the case of the larval stages of the tripod fish *Bathypterois* sp., for which, to the best of our knowledge, we provide the first available photo in environment; see Supplementary Materials, Plate IV). Despite the short period of the investigation conducted in this work, over the annual cycle, mesozooplankton abundance in offshore waters of the Tyrrhenian Sea oscillated within a narrow range and revealed lower seasonal variability [53,54], with peaks occurring in April.

Moonlight has been documented to affect the distribution of plankton biodiversity, with cyclic patterns of vertical movement synchronised with variations of irradiance in all aquatic habitats [55]. Species abundances are higher during new moon and lower during full moon, suggesting that predation pressure from planktivorous species may affect plankton abundance [55]. Moon phase is also used by species to synchronize reproduction [55].

Occurrence of species during the studied period showed two different scenarios according to the moon phase. Whereas the abundance of most taxa [23] seemed not to be directly correlated only with moon cycle, for 12 taxa, there was a clear correlation between their occurrence and moon phases (Supplementary Materials, Figure S3). With the exception of Cestidae, all the other groups of moon-affected taxa clearly avoided full moon. Interestingly, field observation of some of them revealed mating behaviour, such as for Nereids (Supplementary Materials, Plate I), Sepiolidae, and Idioteidae (Supplementary Materials, Plate II), and trophic interactions (Supplementary Materials, Plate III). Juvenile and larval stages of fish were also abundant during the night with a weak moon light, as a direct consequence of lower predation rate [55] (see Supplementary Materials, Plate IV).

4.2. Ecological Insight on the Uncommon Mesopelagic Fish Trachipterus trachypterus

Trachipterus trachypterus is considered rare and with low population densities; however, the mesopelagic habits of the species could lead to an underestimation of its abundance [28]. This fish is only occasionally caught with professional fishing gear, and usually as one or two specimens. Most of the specimens are caught by longline or trawl at depth of 370–700 m [21,24,28,56]; in other cases, they were caught by hand, or observed near the surface or stranded [20,28]. In the present study, the species was observed and caught in exceptionally large numbers. The presence of several juveniles of *T. trachypterus* in the study area could be related to the favorable hydrological and ecological conditions typical of banks and nearby areas. Banks are important areas for fish biodiversity and often host poorly known and rare fish species [57]. Uncommon species, such as *T. trachypterus*, could be good indicators of environmental changes; however, data about the ecology and biology of this species remain scarce.

In ecological communities, it is generally believed that the principal survival option available to marine fish larvae is a fast growth and a quick enhancement of swimming performances [58–60]. The occurrence of *T. trachypterus* displayed a circa-lunar cycle, in contrast with what observed with juveniles of other fish species. A more in depth analysis showed that specimens collected referred to two different size ranges, with the bigger ones (229–157 mm) sampled during almost full moon nights (Supplementary Materials, Table S2, specimens 1 and 2) and the smallerones (91–58 mm) during darkest nights (see also Supplementary Materials, Table S1). Despite the fact that this correlation must be supported by further observations, it is reasonable that the species behaviour changes according to the size, and that it occurs in the surface layer of the pelagic domain as a consequence of hiding or planktotrophic behaviour.

The larval fish phenotype also may have a strong impact on predation through dissembling or active defenses against predators. Diverse and complex morphologies can be found within the same family [61,62], suggesting the occurrence of different predation strategies, even if visual predation by larger fishes is likely the dominant selective force.

Batesian mimicry is a more widespread survival strategy than previously thought, especially within marine communities [63], as demonstrated by the convergence of numbers of distantly related fish taxa on the production of complex features (especially in larval stages and juveniles), like extremely long fin rays (Gadiformes, Lampridae, Lophiidae, Pleuronectiformes, and Serranidae), external guts (Gadiformes, Myctophidae), and even false eyes on stalks (Myctophidae, Stomiidae) [62,63]. Within marine pelagic environments, the evolution of visual Batesian mimicry in larval fishes must meet three fundamental conditions: (1) visual predation must be a strong source of mortality; (2) there must be a relatively common and functional model that the mimic resembles through morphology or behavior, and (3) mimics with only a slight resemblance to the model receives some degree of protection [64,65].

Within this expedition, we had the chance to observe, for the first time, some peculiar behavioural traits of juveniles of *T. trachypterus* which may be linked to Batesian mimicry.

The almost vertical position taken by juveniles of *T. trachypterus* during swimming and the dorsal–fin based locomotion recorded during our study is in agreement with the swimming behaviour previously reported for juveniles of the same species [66]. A similar swimming behaviour was recently reported also for juveniles of *T. arcticus* [67]. On the other hand, new data emerge from our study: 1- the "dorsal/ventral surface exhibition", in which the juvenile fish oriented its dorsal/ventral surface towards the observer; 2- the potential Batesian mimicry. We hypothesise that these two behaviours are related. Dorsal-orienting swimming behaviour may represent a strategy to hide the general shape of the fish and, at the same time, the extension and orientation of caudal and pelvic fins to the potential predator may enhance its appearance as a noxious (or not palatable) organism. The small occasional snap movements of these fins further recall a jellyfish species as a model for the Batesian mimicry (e.g., *Pelagia noctiluca*). Noteworthy, the elongated rays of the anterior part of the dorsal fin and the small orange expansions on these rays are similar in appearance to the cormidia of siphonophores. These peculiar morphological traits seem not to be conserved in the adult and therefore can be considered a functional adaptation to reduce predation risks during the juvenile epipelagic stage of the species. Indeed, this has already been suggested for juveniles and larvae of other mesopelagic fish species, in which particularly long and delicate fins (often in association with specific behaviours and colors) play an important role in reducing mortality due to predation [63,68–70].

5. Conclusions

Despite the numerous investigations of the last decades, the emerging picture of plankton dynamics in the Mediterranean Sea is still largely unexplored. Many factors, such as the structure of the seafloor and continental water inflows, may represent core drivers in the distribution of plankton diversity, structuring complex and dynamic trophic interactions between coastal and pelagic environments. In this research, we show how the vertical transport of coastal nutrients close to the continental shelf boundary sustains a high planktonic diversity and the occurrence of rare mesopelagic fish species. Furthermore, for a total of 12 taxa, a clear correlation between their presence and moon phases is reported.

The concept behind AQUATILIS field-based approach, along with the use of modern filming equipment, allowed us to describe, for the first time, the behaviour of the uncommon mesopelagic fish *T. trachypterus*, depicting the presence of Batesian mimicry in juveniles, and to observe and photograph rare and little known species, such as the larva of *Bathypterois* sp. Our results support the role of field studies as a fundamental instrument improving our knowledge on complex ecosystems, as those inhabited by the mesopelagic communities.

Supplementary Materials: The following are available online at http://www.mdpi.com/2077-1312/8/8/581/s1: Table S1-Planktonic taxa surveyed at the diving site; Figure S1—(a) and (b) show respectively the wind rose of Ponza (2011–2015) and Gaeta (2007-present). Map (c) and (d) show respectively the mean distribution of SST and Salinity from 1987 to present; Figure S2-In the panel (a) is presented in descending order the distribution of SST during 9th–11th of April, during 12th–18th of April and during 19th–28th of April. In the panel (b) the polar scatter of wind indicates the direction (polar orientation) and intensity of wind (x-axis). In the panel (c) is presented in descending order the distribution of Salinity during 9th–11th of April, during 12th–18th of April and during 19th–28th of April; Figure S3-Taxa occurrence according to moon phase; Figure S4. The four sampled juvenile specimens of *Trachipterus trachypterus* from the central Tyrrhenian Sea (A–C), measurements are reported in Table S2; Table S2. Morphometric measurements of the four juvenile specimens of *T. trachypterus* reported in Figure S4 and collected in the central Tyrrhenian Sea; Plate I. Mating behaviour observed during the surveyed period (A) nereids epitokes mating close to the surface; (B) details of epitokes releasing eggs; Plate II. Mating behaviour observed during the surveyed period: (A) Idioteidae on a feather; (B) Sepiolidae; Plate III. Trophic interaction within plankton community, observed during the surveyed period: (A) Scyllaridae phyllosoma feeding on a salp, with detail of its peculiar interaction with jellyfish species as a floating support; (B) *Phylliroe* cfr. *lichtensteinii* feeding on a jellyfish; (C) *Pelagia noctiluca* feeding on a salp; Plate IV. Fish juvenile/larval stages recorded during the expedition: (A) Phycidae (*Phycis* cf. *phycis*); (B) Ipnopidae (*Bathypterois* sp.); (C) Pleuronectiformes; (D) Anguilliformes. Video 1–10. Tracks of behavioural traits of interest. Available online at https://doi.org/10.5281/zenodo.3934365.

Author Contributions: Conceptualization, A.M., A.S., E.B. and F.T.; Methodology, A.M., A.S. and F.T.; Formal Analysis, A.M., F.P.d.M., A.D. and F.T.; Investigation, A.M., A.S. and E.B.; Resources, A.M. and A.S.; Data Curation, A.M., F.P.d.M. and A.D.; Writing—Original Draft Preparation, A.M., F.P.d.M., A.D. and F.T.; Writing—Review & Editing, A.M., A.S., F.P.d.M., A.D., E.B. and F.T.; Project Administration, A.M. and A.S.; Funding Acquisition, A.M. All authors have read and agreed to the published version of the manuscript.

Funding: This research received no external funding.

Acknowledgments: We are keen to thank the AQUATILIS teams and crew, with a special mention for Fedor Bolshakov, Dmitry Ozerov, and Pavel Kremenets. A special thanks to Sviatlana Semenova, for supporting and supervising all the field activities during the expedition. We gratefully acknowledge the AQUATILIS Association for funding the expedition and for the logistic supplies.

Conflicts of Interest: The authors declare no conflict of interest.

References

1. Lévy, M.; Jahn, O.; Dutkiewicz, S.; Follows, M.J.; D'Ovidio, F. The dynamical landscape of marine phytoplankton diversity. *J. R. Soc. Interface* **2015**, *12*, 20150481. [CrossRef]
2. Ferrari, R.; Wunsch, C. Ocean Circulation Kinetic Energy: Reservoirs, Sources, and Sinks. *Annu. Rev. Fluid Mech.* **2009**, *41*, 253–282. [CrossRef]
3. McGillicuddy, D.J. Mechanisms of Physical-Biological-Biogeochemical Interaction at the Oceanic Mesoscale. *Ann. Rev. Mar. Sci.* **2016**, *8*, 125–159. [CrossRef]
4. Mahadevan, A.; Campbell, J.W. Biogeochemical patchiness at the sea surface. *Geophys. Res. Lett.* **2002**, *29*, 31–32. [CrossRef]
5. Lehahn, Y.; D'Ovidio, F.; Koren, I. A Satellite-Based Lagrangian View on Phytoplankton Dynamics. *Ann. Rev. Mar. Sci.* **2018**, *10*, 99–119. [CrossRef] [PubMed]
6. McManus, M.A.; Woodson, C.B. Plankton distribution and ocean dispersal. *J. Exp. Biol.* **2012**, *215*, 1008–1016. [CrossRef] [PubMed]
7. McDonald, L.L. Sampling rare populations. In *Sampling Rare or Elusive Species*; Thompson, W., Ed.; Island Press: Washington, DC, USA, 2004; pp. 11–42.
8. Kelly, R.P.; Port, J.A.; Yamahara, K.M.; Martone, R.G.; Lowell, N.; Thomsen, P.F.; Mach, M.E.; Bennett, M.; Prahler, E.; Caldwell, M.R.; et al. Harnessing DNA to improve environmental management. *Science* **2014**, *344*, 1455–1456. [CrossRef] [PubMed]
9. Wuertz, M. *Mediterranean Pelagic Habitat: Oceanographic and Biological Processes, an Overview*; IUCN: Gland, Switzerland, 2010; ISBN 978-2-8317-1242-0.

10. Bonanno, A.; Zgozi, S.; Basilone, G.; Hamza, M.; Barra, M.; Genovese, S.; Rumolo, P.; Nfate, A.; Elsger, M.; Goncharov, S.; et al. Acoustically detected pelagic fish community in relation to environmental conditions observed in the Central Mediterranean sea: A comparison of Libyan and Sicilian–Maltese coastal areas. *Hydrobiologia* **2015**, *755*, 209–224. [CrossRef]
11. Giannoulaki, M.; Iglesias, M.; Tugores, M.P.; Bonanno, A.; Patti, B.; De Felice, A.; Leonori, I.; Bigot, J.L.; Tičina, V.; Pyrounaki, M.M.; et al. Characterizing the potential habitat of European anchovy *Engraulis encrasicolus* in the Mediterranean Sea, at different life stages. *Fish. Oceanogr.* **2013**, *22*, 69–89. [CrossRef]
12. Tugores, P.; Giannoulaki, M.; Iglesias, M.; Bonanno, A.; Tičina, V.; Leonori, I.; MacHias, A.; Tsagarakis, K.; Díaz, N.; Giráldez, A.; et al. Habitat suitability modelling for sardine *Sardina pilchardus* in a highly diverse ecosystem: The Mediterranean Sea. *Mar. Ecol. Prog. Ser.* **2011**, *443*, 181–205. [CrossRef]
13. Damgaard, C.; Weiner, J. It's About Time: A Critique of Macroecological Inferences Concerning Plant Competition. *Trends Ecol. Evol.* **2017**, *32*, 86–87. [CrossRef] [PubMed]
14. Watson, D.L.; Harvey, E.S.; Fitzpatrick, B.M.; Langlois, T.J.; Shedrawi, G. Assessing reef fish assemblage structure: How do different stereo-video techniques compare? *Mar. Biol.* **2010**, *157*, 1237–1250. [CrossRef]
15. Olney, J.E.; McBride, R.S. Intraspecific variation in batch fecundity of American shad: Revisiting the paradigm of reciprocal latitudinal trends in reproductive traits. *Am. Fish. Soc. Symp.* **2003**, *2003*, 185–192.
16. Walters, V.; Fitch, J. The families and genera of the lampridiform (Allotriognath) suborder Trachipteroidei. *Calif. Fish Game* **1960**, *46*, 441–451.
17. Froese, R.; Pauly, D. FishBase. Available online: www.fishbase.org (accessed on 1 May 2018).
18. Dulčić, J. First record of ribbon fish larva, *Trachipterus trachypterus*, from the Eastern Adriatic. *Cybium* **1996**, *20*, 101–102.
19. Dulčić, J. First record of scalloped ribbon fish, *Zu cristatus* (Pisces: Trachipteridae), eggs in the Adriatic Sea. *J. Plankton Res.* **2002**, *24*, 1245–1246. [CrossRef]
20. Borme, D.; Voltolina, F. On the occurrence of ribbon fish *Trachipterus trachypterus* (Gmelin, 1789) in the Gulf of Trieste (northern Adriatic Sea). *Ann. Ser. Historia Nat.* **2006**, *16*, 181.
21. Psomadakis, P.N.; Scacco, U.; Vacchi, M. Recent findings of some uncommon fishes from the central Tyrrhenian Sea. *Cybium* **2006**, *30*, 297–304.
22. Psomadakis, P.N.; Bottaro, M.; Vacchi, M. On two large specimens of *Zu cristatus* (Trachipteridae) from the Gulf of Genoa (NW Mediterranean). *Cybium* **2007**, *31*, 480–482.
23. Bradai, M.N.; El Ouaer, A. New record of the scalloped ribbon fish, *Zu cristatus* (Osteichthyes: Trachipteridae) in Tunisian waters (central Mediterranean). *Mar. Biodivers. Rec.* **2014**, *5*, e59. [CrossRef]
24. Mytilineou, C.; Anastasopoulou, A.; Christides, G.; Bekas, P.; Smith, C.J.; Papadopoulou, K.N.; Lefkaditou, E.; Kavadas, S. New records of rare deep-water fish species in the Eastern Ionian Sea (Mediterranean Sea). *J. Nat. Hist.* **2013**, *47*, 1645–1662. [CrossRef]
25. Bono, G.; Gancitano, S.; Bosch-Belmar, M.; Giusto, G.B.; Okpala, C.O.R.; Scannella, D.; Geraci, M.L.; Falsone, F. Occurrence of two rare species from order Lampriformes: Crestfish *Lophotus lacepede* (Giorna, 1809) and scalloped ribbonfish *Zu cristatus*(Bonelli, 1819) in the northern coast of Sicily, ItalyPojava dviju rijetkih vrsta iz reda Lampriformes: B. *Acta Adriat.* **2017**, *58*, 137–146. [CrossRef]
26. Bianco, P.G.; Zupo, V.; Ketmaier, V. Occurrence of the scalloped ribbonfish *Zu cristatus* (Lampridiformes) in coastal waters of the central Tyrrhenian Sea, Italy. *J. Fish Biol.* **2006**, *68*, 150–155. [CrossRef]
27. Crocetta, F.; Agius, D.; Balistreri, P.; Bariche, M.; Bayhan, Y.K.; Çakir, M.; Ciriaco, S.; Corsini-Foka, M.; Deidun, A.; El Zrelli, R.; et al. New mediterranean biodiversity records (October 2015). *Mediterr. Mar. Sci.* **2015**, *16*, 682–702. [CrossRef]
28. Tiralongo, F.; Lillo, A.O.; Tibullo, D.; Tondo, E.; Martire, C.L.; D'Agnese, R.; Macali, A.; Mancini, E.; Giovos, I.; Coco, S.; et al. Monitoring uncommon and non-indigenous fishes in Italian waters: One year of results for the AlienFish project. *Reg. Stud. Mar. Sci.* **2019**, *28*, 100606. [CrossRef]
29. Paladini de Mendoza, F.; Bonamano, S.; Martellucci, R.; Melchiorri, C.; Consalvi, N.; Piermattei, V.; Marcelli, M. Circulation during storms and dynamics of suspended matter in a sheltered coastal area. *Remote Sens.* **2018**, *10*, 602. [CrossRef]
30. Martellucci, R.; Pierattini, A.; Paladini de Mendoza, F.; Melchiorri, C.; Piermattei, V.; Marcelli, M. Physical and biological water column observations during summer sea/land breeze winds in the coastal northern Tyrrhenian Sea. *Water (Switzerland)* **2018**, *10*, 1673. [CrossRef]

31. Kronfeld-Schor, N.; Dominoni, D.; de la Iglesia, H.; Levy, O.; Herzog, E.D.; Dayan, T.; Helfrich-Forster, C. Chronobiology by moonlight. *Proc. R. Soc. B Biol. Sci.* **2013**, *280*, 20123088. [CrossRef]
32. Last, K.S.; Hobbs, L.; Berge, J.; Brierley, A.S.; Cottier, F. Moonlight Drives Ocean-Scale Mass Vertical Migration of Zooplankton during the Arctic Winter. *Curr. Biol.* **2016**, *26*, 244–251. [CrossRef]
33. Benoit-Bird, K.J.; Au, W.W.L.; Wisdom, D.W. Nocturnal light and lunar cycle effects on diel migration of micronekton. *Limnol. Oceanogr.* **2009**, *54*, 1789–1800. [CrossRef]
34. Drazen, J.C.; De Forest, L.G.; Domokos, R. Micronekton abundance and biomass in Hawaiian waters as influenced by seamounts, eddies, and the moon. *Deep Sea Res. Part I Oceanogr. Res. Pap.* **2011**, *58*, 557–566. [CrossRef]
35. Rubolini, D.; Maggini, I.; Ambrosini, R.; Imperio, S.; Paiva, V.H.; Gaibani, G.; Saino, N.; Cecere, J.G. The Effect of Moonlight on Scopoli's Shearwater *Calonectris diomedea* Colony Attendance Patterns and Nocturnal Foraging: A Test of the Foraging Efficiency Hypothesis. *Ethology* **2015**, *121*, 284–299. [CrossRef]
36. Aksnes, D.L.; Røstad, A.; Kaartvedt, S.; Martinez, U.; Duarte, C.M.; Irigoien, X. Light penetration structures the deep acoustic scattering layers in the global ocean. *Sci. Adv.* **2017**, *3*, e1602468. [CrossRef] [PubMed]
37. Ochoa, J.; Maske, H.; Sheinbaum, J.; Candela, J. Diel and lunar cycles of vertical migration extending to below 1000 m in the ocean and the vertical connectivity of depth-tiered populations. *Limnol. Oceanogr.* **2013**, *58*, 1207–1214. [CrossRef]
38. Anderson, M.J. Permutation tests for univariate or multivariate analysis of variance and regression. *Can. J. Fish. Aquat. Sci.* **2001**, *58*, 626–639. [CrossRef]
39. Anderson, M.J.; Gorley, R.N.; Clarke, K.R. *PERMANOVA+ for PRIMER: Guide to Software and Statistical Methods*; PRIMER-E: Plymouth, UK, 2008; pp. 1–214.
40. Villegas-Zurita, F.; Castillejos-Moguel, F.; Benítez-Villalobos, F.; Urbán-Ramírez, J. Alpha diversity of marine mammals of the Mexican South Pacific. *Rev. Mex. Biodivers.* **2018**, *89*, 898–909. [CrossRef]
41. Morris, C. Multivariate Analysis of Ecological Data Using Canoco 5. *Afr. J. Range Forage Sci.* **2015**, *32*, 289–290. [CrossRef]
42. Bakun, A.; Agostini, V.N. Seasonal patterns of wind-induced upwelling/downwelling in the Mediterranean Sea. *Sci. Mar.* **2001**, *65*, 243–257. [CrossRef]
43. Haury, L.R.; McGowan, J.A.; Wiebe, P.H. Patterns and Processes in the Time-Space Scales of Plankton Distributions. In *Spatial Pattern in Plankton Communities*; Springer: New York, NY, USA, 1978; pp. 277–327.
44. Le Fèvre, J. Aspects of the Biology of Frontal Systems. *Adv. Mar. Biol.* **1987**, *23*, 163–299. [CrossRef]
45. Joyce, T.M. Varieties of ocean fronts. In *Baroclinic Instability and Ocean Fronts*; Technical Report No. 83-41; Woods Hole Oceanographic Institution: Falmouth, MA, USA, 1983; Volume 1, p. 59.
46. Belkin, I.M.; Cornillon, P.C.; Sherman, K. Fronts in Large Marine Ecosystems. *Prog. Oceanogr.* **2009**, *81*, 223–236. [CrossRef]
47. Franks, P.J.S. Phytoplankton blooms at fronts: Patterns, scales, and physical forcing mechanisms. *Rev. Aquat. Sci.* **1992**, *6*, 121–137.
48. Genin, A.; Jaffe, J.S.; Reef, R.; Richter, C.; Franks, P.J.S. Swimming against the flow: A mechanism of zooplankton aggregation. *Science* **2005**, *308*, 860–862. [CrossRef] [PubMed]
49. Bost, C.A.; Cotté, C.; Bailleul, F.; Cherel, Y.; Charrassin, J.B.; Guinet, C.; Ainley, D.G.; Weimerskirch, H. The importance of oceanographic fronts to marine birds and mammals of the southern oceans. *J. Mar. Syst.* **2009**, *78*, 363–376. [CrossRef]
50. Bignami, F.; Böhm, E.; D'Acunzo, E.; D'Archino, R.; Salusti, E. On the dynamics of surface cold filaments in the Mediterranean Sea. *J. Mar. Syst.* **2008**, *74*, 429–442. [CrossRef]
51. Wang, D.-P.; Vieira, M.E.C.; Salat, J.; Tintoré, J.; La Violette, P.E. A shelf/slope frontal filament off the northeast Spanish Coast. *J. Mar. Res.* **2008**, *46*, 321–332. [CrossRef]
52. Siokou-Frangou, I.; Christaki, U.; Mazzocchi, M.G.; Montresor, M.; Ribera D'Alcala, M.; Vaque, D.; Zingone, A. Plankton in the open mediterranean Sea: A review. *Biogeosciences* **2010**, *7*, 1543–1586. [CrossRef]
53. Di Carlo, B.S.; Ianora, A.; Fresi, E.; Hure, J. Vertical zonation patterns for mediterranean copepods from the surface to 3000 m at a fixed station in the Tyrrhenian Sea. *J. Plankton Res.* **1984**, *6*, 1031–1056. [CrossRef]
54. de Puelles, M.L.F.; Pinot, J.-M.; Valencia, J. Variabilite saisonniere et interannuelle du zooplancton au large de Majorque (mer des Baleares, Mediterranee occidentale). *Oceanol. Acta* **2003**, *5*, 673–686.
55. Chakraborty, U. Effects of different phases of the lunar month on living organisms. *Biol. Rhythm Res.* **2020**, *51*, 254–282. [CrossRef]

56. Farias, I.; Moura, T.; Figueiredo, I.; Vieira, A.R.; Serra-Pereira, B.; Serrano Gordo, L. Northernmost occurrence of the ribbonfish *Trachipterus trachypterus* (Gmelin, 1789) in the NE Atlantic: The Portuguese continental shelf: Short communication. *J. Appl. Ichthyol.* **2010**, *26*, 143–144. [CrossRef]
57. Consoli, P.; Esposito, V.; Battaglia, P.; Altobelli, C.; Perzia, P.; Romeo, T.; Canese, S.; Andaloro, F. Fish distribution and habitat complexity on banks of the strait of sicily (central mediterranean sea) from Remotely-Operated Vehicle (ROV) explorations. *PLoS ONE* **2016**, *11*, e0167809. [CrossRef] [PubMed]
58. Houde, E.D. Fish Early Life Dynamics and Recruitment Variability. In Proceedings of the American Fisheries Society Symposium, 10 Annual Larval Fish Conference, Miami, FL, USA, 18–23 May 1987.
59. Anderson, J.T. A Review of Size Dependant Survival during Pre-recruit Stages of Fishes in Relation to Recruitment. *J. Northwest Atl. Fish. Sci.* **1988**, *8*, 55–66. [CrossRef]
60. Hare, J.A.; Cowen, R.K. Size, growth, development, and survival of the planktonic larvae of Pomatomus saltatrix (Pisces: Pomatomidae). *Ecology* **1997**, *78*, 2415–2431. [CrossRef]
61. Wells, K.D. The Origin and Evolution of Larval Forms. *Copeia* **2000**. [CrossRef]
62. Miller, B.S.; Kendall, A.W. *Early Life History of Marine Fishes*; University of California Press: Berkeley, CA, USA, 2009; ISBN 9780520249721.
63. Greer, A.T.; Woodson, C.B.; Guigand, C.M.; Cowen, R.K. Larval fishes utilize Batesian mimicry as a survival strategy in the plankton. *Mar. Ecol. Prog. Ser.* **2016**, *551*, 1–12. [CrossRef]
64. Huheey, J.E. Mathematical Models of Mimicry. *Am. Nat.* **1988**, *131*, S22–S41. [CrossRef]
65. Caley, M.J.; Schluter, D. Predators favour mimicry in a tropical reef fish. *Proc. R. Soc. B Biol. Sci.* **2003**, *270*, 667–672. [CrossRef]
66. Nishimura, S.; Hirosaki, Y. Observations on the swimming behavior of some taeniosomous fishes in aquaria and in nature. *Publ. Seto Mar. Biol. Lab.* **1964**, *12*, 165–171. [CrossRef]
67. Moritz, T.; Stümer, D.; Jakobsen, K.; Jakobsen, J. Observations on two live specimens of *Trachipterus arcticus* (Lampriformes: Trachipteridae) from the Azores. *Cybium* **2015**, *39*, 78–80.
68. Sazima, I. Juvenile grunt (Haemulidae) mimicking a venomous leatherjacket (Carangidae), with a summary of Batesian mimicry in marine fishes. *J. Ichthyol. Aquat. Biol.* **2002**, *6*, 61–68.
69. Reininger, M. Mimicry in juvenile wrasses: Ecological and behavioural aspects of a coris-amphiprion relationship in the northern red sea. *Zool. Middle East* **2011**, *54*, 23–34. [CrossRef]
70. Tan, H.H. Apparent mimicry of jellyfish by juvenile pomfret, *Pampus chinensis* (Teleostei: Stromateidae). *Nat. Singap.* **2008**, *1*, 139–142.

Journal of
Marine Science and Engineering

MDPI

Article

The Challenge to Observe Antarctic Toothfish (*Dissostichus mawsoni*) under Fast Ice

Davide Di Blasi [1,2,*], Simonepietro Canese [3,4], Erica Carlig [1], Steven J. Parker [5], Eva Pisano [1], Marino Vacchi [1] and Laura Ghigliotti [1]

1 Institute for the Study of Anthropic Impacts and Sustainability in Marine Environment–National Research Council, via De Marini 6, 16149 Genoa, Italy; erica.carlig@ias.cnr.it (E.C.); eva.pisano@ias.cnr.it (E.P.); marino.vacchi@ias.cnr.it (M.V.); laura.ghigliotti@cnr.it (L.G.)
2 Department of Earth, Environmental and Life Sciences, University of Genoa, 16149 Genoa, Italy
3 Stazione Zoologica Anthon Dohrn, 80100 Naples, Italy; simonepietro.canese@szn.it
4 Italian Institute for Environmental Protection and Research, 00185 Rome, Italy
5 National Institute of Water and Atmospheric Research Ltd, 7010 Nelson, New Zealand; steve.parker@niwa.co.nz
* Correspondence: davide.diblasi@ias.cnr.it

Abstract: In situ observation of Antarctic toothfish (*Dissostichus mawsoni*) is challenging as they typically live at depths greater than 500 m, in dark and ice-covered Antarctic waters. Searching for adequate methodologies to survey Antarctic toothfish in their habitat, we tested a miniaturized Baited Remote Underwater Video camera (BRUV), deployed through holes drilled in the sea ice in the Ross Sea region, over three field seasons. In 2015 three BRUVs were deployed at McMurdo Sound, and paired with a vertical longline sampling. In 2017, three opportunistic deployments were performed at Terra Nova Bay. In 2018 seven deployments at Terra Nova Bay provided preliminary data on the habitat preferences of the species. The design and configuration of the mini-BRUV allowed to collect high-quality video imagery of 60 Antarctic toothfish in 13 deployments from the fast sea ice. The behaviour of fish at the bait, intra-species interactions, and potential biases in individual counting were investigated, setting baselines for future studies on the abundance and distribution of Antarctic toothfish in sea-ice covered areas. This work represents the first step towards the development of protocols for non-extractive monitoring of the Antarctic toothfish in the high-Antarctica coastal shelf areas, of great value in the Ross Sea region where the largest MPA of the world has recently been established.

Keywords: BRUV; Ross Sea; video sampling; Antarctica

Citation: Di Blasi, D.; Canese, S.; Carlig, E.; Parker, S.J.; Pisano, E.; Vacchi, M.; Ghigliotti, L. The Challenge to Observe Antarctic Toothfish (*Dissostichus mawsoni*) under Fast Ice. *J. Mar. Sci. Eng.* **2021**, *9*, 255. https://doi.org/10.3390/jmse9030255

Academic Editor: Francesco Tiralongo

Received: 23 January 2021
Accepted: 25 February 2021
Published: 28 February 2021

1. Introduction

The Antarctic toothfish (*Dissostichus mawsoni*) is the largest notothenioid fish inhabiting Antarctic continental waters, where it is a keystone species in the food web as a high-trophic-level predator. Since 1998, this species has been targeted by commercial fisheries in the Ross Sea (conventionally defined by the 60° S parallel, 150° E and 150° W meridians, and the corresponding coastline of Antarctica) managed by the Commission for the Conservation of Antarctic Marine Living Resources (CCAMLR), according to an ecosystem-based precautionary approach (www.ccamlr.org, accessed on 24 February 2021). Such an approach requires broad understanding of the species' life history and ecology as foundation for management [1]. For Antarctic toothfish, biological and ecological information has mostly been collected by observers onboard of commercial fishery vessels in offshore, deeper, and ice-free waters [2,3].

In the Ross Sea region, a large Antarctic toothfish population has been identified based on genetics and mark-recapture studies [2]. The population spans a wide geographic range from the spawning habitats in northern areas of the Pacific Antarctic Ridge to the feeding grounds on the continental slope, and juvenile habitats in the deeper zones of the continental shelf [3–5]. However, while data on size, distribution, diet and reproductive status of Antarctic toothfish from vessel-based surveys exists, it does not cover the full extent or ecological niche of the species seasonally or spatially, as vessel-based data is confined to ice free waters in summer and autumn.

Since 2018, with the implementation of the Ross Sea region Marine Protected Area (Ross Sea region MPA, CCAMLR Conservation Measure 91-05), the continental shelf has been closed to commercial fishing, further limiting information from this area, and constraining the ability to monitor the effect of the MPA. While coastal areas of the Ross Sea shelf are the main locations where toothfish are preyed upon by their main predators, Weddell seals (*Leptonychotes weddellii*) and Killer Whales (*Orcinus orca*), these fast-ice covered areas are neglected by monitoring, and few information is currently available on the Antarctic toothfish population dynamics in those areas [6–10]. Surveying through the sea ice requires large holes made through ice more than 2-m thick to take out large toothfish [8]. This is time consuming and requires the use of large and heavy equipment, both characteristics making such an extractive methodology logistically demanding in high Antarctica and limiting survey activities over large areas. To enable monitoring of Antarctic toothfish in fast-ice-covered shelf areas, innovative methods need to be developed to overcome logistical constraints, including non-extractive methods for working in the MPA.

Non-extractive methods for the study of marine fauna include acoustic and visual techniques. Underwater acoustics is largely used for studying zooplankton as well as pelagic species, but it is not effective for organisms that reside or move close to the bottom [11], such as the Antarctic toothfish, as the vertical resolution, especially at the appropriate depths, can obscure several metres of demersal habitat, and targets need to be identified from acoustic characteristics [12,13]. Conversely, underwater video techniques allow to record abundance and distribution of target species both in the water column and close to the bottom, where targets can be visually identified. Additional benefit of video sampling is the ability to observe behaviour, habitat association as well as intra- and inter-species relationships (see [11] for a review).

Among the video techniques, the Baited Remote Underwater Video systems (BRUVs) methodology is conceptually simple and based on a recording video camera that documents the arrival of organisms attracted to a baited lander [14,15]. Such a technique, which minimizes observer biases and gear selectivity associated with other survey methods, is likely appropriate for fish such as the Antarctic toothfish, characterized by good olfactory capabilities [16] and with benthic scavenger feeding habits [17]. BRUVs allow video documentation of species presence, size, and behaviour (e.g., swimming speed, feeding mode, searching). Relative abundance metrics can also be developed [18], provided that the bias associated with counting individuals that enter the field of view multiple times is accounted for [19].

Since the mid-1990s, BRUVs have been used in temperate, tropical and subtropical areas, mostly to assess the effect of Marine Protected Areas, document species behaviour, or assess changes in fish assemblages [11,20,21]. In polar waters, BRUVs have been less employed so far. BRUVs were used in the marine waters of the northern Canadian territory of Nunavut [22,23]. Very few baited camera deployments were performed in the Southern Ocean prior to present work, none in high Antarctica. An autonomous lander was deployed around South Georgia and Falkland Island to estimate the abundance of the congeneric Patagonian toothfish (*Dissosticus eleginoides*) independently from the fishery catch data [24]. A BRUV was set by SCUBA diver in shallow waters at Adelaide Island (West Antarctic Peninsula) to study the response of scavengers to feeding cues in the area [25]. Within Ryder Bay, in the West Antarctic Peninsula, a baited camera system was used to examine the distribution of scavenging fauna in relation to the spatial variation in exposure to iceberg scouring [26]. In those cases, the BRUVs were large in size and/or needed to be set underwater by a SCUBA diver, both characteristics unsuitable for work in sea-ice covered areas and at over 500 m depth.

Here, we conducted a trial of the feasibility, efficacy and reliability of a mini BRUV to study the Antarctic toothfish in fast ice-covered shelf areas. In the frame of collaborative researches between New Zealand and Italy, a pilot study was undertaken during three Antarctic field seasons, at two different locations (nearby-located to the New Zealand and the Italian Antarctic research stations that gave logistic support to make the study possible) within the Western Ross Sea region. The study aimed to (i) investigate the Antarctic toothfish behaviour at the bait and evaluate its potential influence on the calculation of relative abundance metrics, (ii) test the reliability of the results obtained by BRUV by comparison with those collected by extractive techniques, and (iii) set baselines for future studies on the distribution and abundance of the species in shelf areas.

The work promotes the diffusion of such a non-extractive technique for monitoring and sampling within the Ross Sea region MPA and sets the bases for use of BRUVs in other areas around the Antarctic continent.

2. Materials and Methods

2.1. Study Area and Sampling

The work was part of New Zealand-Italian collaborative activities conducted at McMurdo Sound, near the New Zealand's Scott Base in November 2015, and Terra Nova Bay, near the Italian Mario Zucchelli Station in November 2017 and 2018.

2.1.1. General Description of the Study Area

The fast-ice study area is located in the western Ross Sea (Figure 1A) with a mean depth of about 500 meters and bathymetry associated with local volcanos and scouring by glacial ice. The shape of the seafloor and depth direct tidal currents comprised Ice Shelf Water (ISW) and the water produced beneath the freezing sea ice (Ross Sea Shelf Water, RSSW). In the spring months, currents are usually less than 10 cm s^{-1} and flow in a north-south direction [27].

All deployments were made through the 1.5–2.5 m thick fast sea ice that covers the sea surface for 9–10 months a year, from March until January. The holes extended through the platelet ice layer, a feature unique to coastal Antarctic zones [28,29]. Two locations were sampled, McMurdo Sound and Terra Nova Bay (Figure 1B). The former is a long depression up to 1000 m deep extending from the Ross Ice Shelf and bordering Ross Island [27]. The latter is along the Victoria Land coast and is a steep seafloor consisting of granitic rock ridges and gullies filled with gravel, clay, and silt [27,30]. Deployments in Terra Nova Bay were performed in Silverfish Bay which, within the vast Ross Sea region MPA, also constitutes an Antarctic Specially Protected Area (ASPA 173) of the Antarctic Treaty.

2.1.2. Mini-BRUV

One of the main points for consideration when working in Antarctica is logistics. With sampling sites spanning over a large area, the constraints in the transportation, included those deriving by the use of helicopters, need to be considered. Accordingly, the BRUV system was designed to be light and portable to optimize loads during transport. Furthermore, the size of the BRUV was minimized in order to allow deployments from relatively small holes in the sea ice, thus avoiding carrying around the large and heavy equipment that would be necessary to drill large holes and saving time.

The mini-BRUV system consisted of two cylindrical housings made of Delrin® POM (DuPont™) of 70 mm external diameter and 250 mm in length. At the end of one cylinder was a 15 mm thick, flat acrylic camera and light sensor port. The other end contained a flat Delrin plate with a Seacon Electrical Wet-Mate bulkhead connector. The cylinder held a full HD Mobius camera (with a 64 Gb memory card and an Arduino Micro microcontroller board), and a NiMH battery pack. The other cylinder of the same size held a MR16 LED lamp (6 Watt, 12 Volt, cool white), and a dedicated NiMH battery pack. The two cylinders were fixed in parallel and connected by cable (Figure 2).

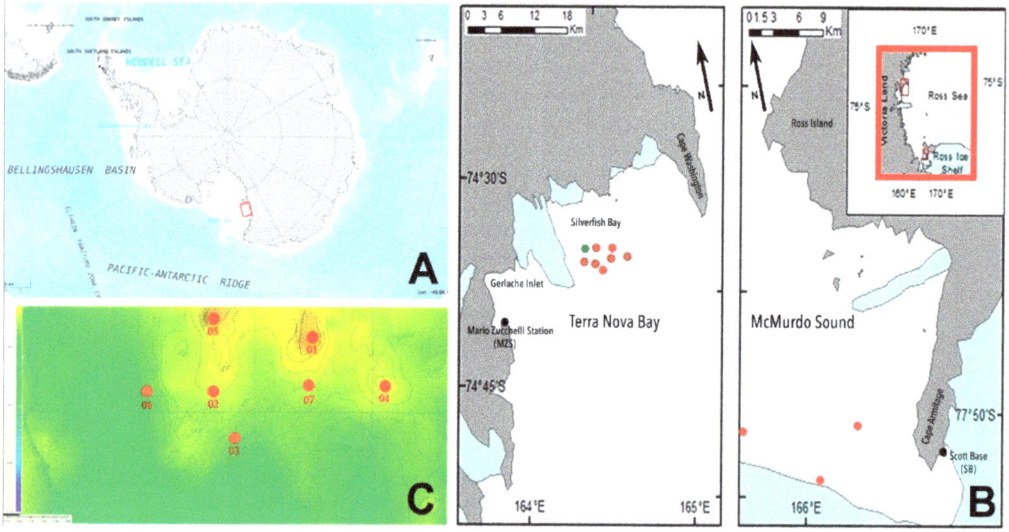

Figure 1. The study area. (**A**) Location of the study area within Antarctica continental shelf (red frame); (**B**) Sampling sites at Terra Nova Bay (left) and McMurdo Sound (right). Ice tongues or ice shelves are marked in light-blue colour; lands are marked in grey; white colour corresponds to seawater, partly covered by sea ice during the sampling periods. The sampling stations are in areas covered by fast ice in the late spring. Red dots indicate the sites sampled in 2018 at Terra Nova Bay (left) and in 2015 at McMurdo Sound (right). The green dot corresponds to the sampling site at which the BRUV was opportunistically deployed in 2017, three times with the bait and two times without the bait. (**C**) Bottom geomorphology at Silverfish Bay (Terra Nova Bay), the red dots are the seven sampling sites where Baited Remote Underwater Video camera (BRUV) deployments were carried out in 2018.

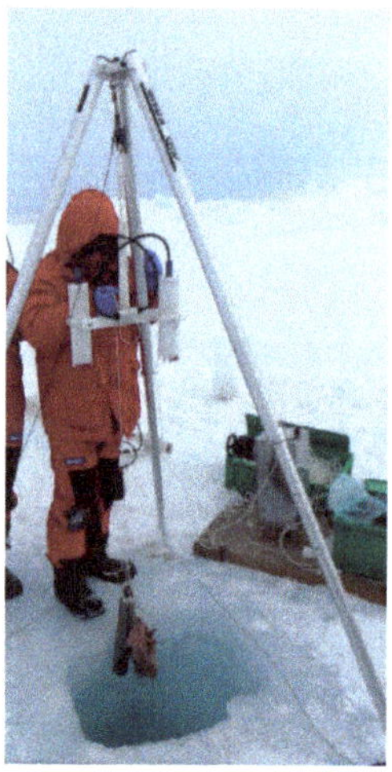

Figure 2. Essential structure of the mini-BRUV system. The prototype was deployed through ice holes of about 40 cm diameter.

The Arduino microcontroller was programmed to switch the camera and light "on" or "off". During the first year of field activity, at McMurdo Sound, the camera was set to perform one minute of registration and one minute of pause to optimize the battery duration. Given the good performance of the batteries in this first trial, the protocol was adjusted during the field activities at Terra Nova Bay, where registration was continuous.

Approximately 1 kg of squid (*Notodarus gouldi*), routinely used in the Antarctic toothfish longline fishery, was used as the bait. It was fixed 2 m below the camera, just above the 8 kg clump weight. Three 150 mm diameter trawl floats (800 g buoyancy each) were placed above the mini-BRUV to adjust buoyancy and suspend it 2 m above the clump weight. The camera configuration was vertical with the view towards the seafloor, resulting in a field of view of about 2 m in diameter. The system was slowly lowered with a 3 mm nylon rope to the seafloor by hand or with a winch.

2.1.3. Sampling Design

The study included the following three lines of investigation, covering major aspects related to the use of BRUV systems to monitor Antarctic toothfish in ice-covered areas.

1. Toothfish behaviour at the bait

Three BRUV deployments were performed at McMurdo Sound (2015) and ten at Silverfish Baywithin Terra Nova Bay (three at the same station in 2017, and seven at different stations in 2018) (Figure 1B,C). In addition, in 2017, the mini-BRUV was deployed twice at Silverfish Bay without bait, as negative control to evaluate the attractive effect of the bait.

The residence time, how long on average a specimen remains in the vicinity of the bait (from first appearance to last appearance in the case of individuals that keep coming back into the field of the camera), was calculated in the 10 continuous videos performed at Terra Nova Bay during 2017 and 2018. Behaviour at the bait was evaluated as potential source of bias in the count of toothfish by BRUV [19]. When more than one individual was present in the camera field of view, the behaviour of each one was annotated and classified as "neutral", "agonistic aggressor" or "agonistic subordinate". For each agonistic event, the behaviour of both the aggressor and the subordinate fish were noted and described as follows: "stay", when the fish did not appear altered by the event; "weak reaction", when the fish reacted with rapid movements and appeared disturbed or when it escaped but returned under the field of view after a few seconds; "escape", when the fish moved away and did not return in proximity of the bait. The size of the individuals involved in the different events was included in the analysis.

2. Comparison between BRUV data and longline catch

This activity was implemented at McMurdo Sound in 2015, where a scientific random stratified vertical longline survey was being conducted from the sea ice, as part of a large-scale monitoring programme [10]. The aim of this first field activity was to perform preliminary tests on the reliability of the BRUV systems in collecting Antarctic toothfish abundance data. To this end, three longline fishing stations (namely Station 20, 28, and 29, details in [10]) were considered. At each station, vertical longlines, armed with 15/0 size hooks (EZ-baiter, Mustad) baited with squid, were set three times for approximately 18 h each. The mini-BRUV was deployed at the three longline fishing stations, once per each station. The camera deployments were opportunistically performed after the retrieval of one of the longlines and before the deployment of the subsequent longline set, thus resulting in BRUV soak times ranging between 19 and 24 h. The number of individuals recorded by BRUV in a determined time of unit was compared to the vertical longline catch, expressed as average catch obtained from three replicate fishing events per station.

3. Habitats and distribution of Antarctic toothfish individuals

A preliminary study, aimed at investigating the relation between physical characteristics of the sea bottom, depth and presence of Antarctic toothfish, was performed in 2018 at Terra Nova Bay. The study included BRUV deployments at seven stations in Silverfish Bay. The stations were representative of different geomorphological features (ridge and trench) and falling in a range of bottom depths from about 200 m to more than 500 m. Soak time ranged between 1 to 6 h, depending on the contingencies of the time allowed for field activity and the weather. Information on relief, substrate type, and benthos coverage of the seafloor was recorded for each station. Depth was measured by echosounder during the video recording. The substrate type and the coverage by benthic organisms were evaluated through image analysis of the sea floor. Sediment granularity was not easily distinguishable from video footages. Therefore, we classified the substrate as "soft" when the seafloor was homogenously composed by clay, silt or sand, and "mixed" when rocky formations were visible. The benthos coverage was defined as percentage of seafloor surface in the field of view of the camera covered by epibenthic organisms. Examples of seafloor features are shown in Figure 3.

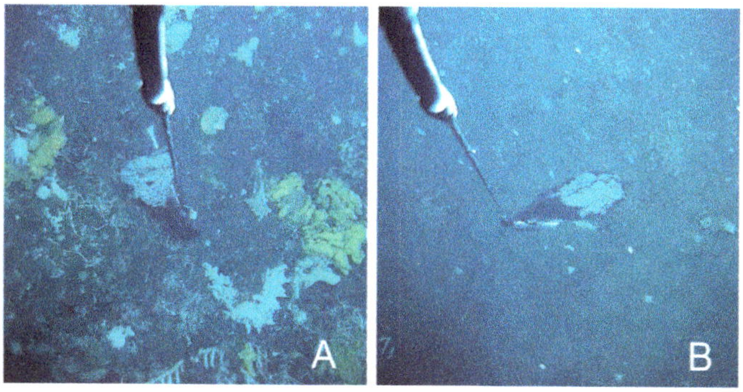

Figure 3. Examples of seafloor features. (**A**) Mixed substrate with high (90%) benthos coverage (Terra Nova Bay, site 05). (**B**) Soft substrate with low (10%) benthos coverage (Terra Nova Bay, site 06).

2.2. Video Analysis

Videos were screened in full with VLC Media Player 3.0.5 Vetinari Software. Identification of individual Antarctic toothfish was performed by extracting frames each time a toothfish was within the field of view and establishing individual and unambiguous key features such as colour patterns, parasites, scars, or other recognizable marks (Figure 4).

Figure 4. Examples of identifying marks on toothfish individuals. (**A**) Distinctive colour pattern of the pectoral fins and presence of a parasite on the left side of the trunk. (**B**) Dark spot pattern along the body.

ImageJ software was used to estimate the total length. For each individual identified, the length of the fish was calculated by comparison with the cylindrical weight on the seafloor (33 cm long) when they were both at a similar distance from the camera. In order to minimize the effect of potential errors in the measurements, we categorised the fish length into three length classes following [3], corresponding to immature (L < 100 cm), maturing (L = 100–130 cm), and mature fish (L > 130 cm).

2.3. Abundance Indices

Relative abundance metrics, typically calculated from BRUV footages, were derived from videos recorded in 2018 at Terra Nova Bay.

The Catch Per Unit of Effort (CPUE) is an index widely used in fishery research. It was adapted to the BRUV analyses [31,32] and calculated as the total number of fish recorded, divided by the time of the video recording, and expressed as number of individual fish observed per hour. CPUE is not widely used in BRUV analysis, because unbiased value requires to identify each individual fish that enter the field of view of the camera during the entire observation period. This is especially difficult where large numbers of individuals are present, as it is often the case in shallow tropical or temperate waters, or for species whose individuals are not easily distinguishable. Given the relatively low densities of toothfish, and the possibility to identify individuals, potential bias in the use of CPUE is limited; therefore, we decided to consider this metrics.

The Mean Number (MeanN) [33], commonly used for dense and/or multispecies shoals of fish [34–44], was calculated from the maximum number of fish in a single frame from the whole video (Maximum Number, MaxN). While the advantage of this metric is to avoid recounts of same individuals [45] without the necessity to identify each individual, it can under-estimate the true abundance of fish visiting the bait, given that only a portion of the fish contribute to MaxN [46]. Therefore, we decided to consider also the MeanN, that is the average of the mean MaxN from 1-hour periods throughout the duration of the recording [33]. For the calculation of MeanN, final segments videos shorter than 1 h were discarded.

The Time of First Arrival (TFA) [47,48] was calculated as the time in minutes that passed from when the BRUV reached the seafloor and the first record of an Antarctic toothfish that entered in the field of view of the camera.

The correlation between metrics was tested through Pearson's correlation coefficient, and the relative abundance of Antarctic toothfish deriving from the various metrics was considered in the frame of a suite of environmental drivers (geomorphological features, depth, substrate type, and benthos coverage).

3. Results

3.1. Fish Behaviour

A total of 60 toothfish were observed during the 13 deployments carried out between McMurdo Sound and Terra Nova Bay. The residence time of individuals around the bait, calculated for 52 individuals during the continuous video recordings at Terra Nova Bay, was highly variable with mean residence time of 8 min 24 s ± 9 min 36 s SD (maximum 44 min, minimum 10 s). No toothfish were recorded when the BRUV was deployed without the bait. Of the observed fish, 44 were recorded close to the bait together with at least another individual in the same frame (Video S1). Most toothfish (25 individuals) were neutral to the presence of other individuals within the field of view; two individuals left the field of view without any intra-specific interaction; one was apparently involuntarily bitten by another individual and had a weak escape reaction, swimming away and returning a few seconds later; two fish increased the swimming speed in the field of view after inadvertent contact; and in 13 cases evident agonistic behaviour were recorded (Figure 5).

A descriptive analysis was developed on the 13 cases that showed agonistic events from the combination of the type of event, size classes involved, and subsequent reactions (Figure 6). In 5 cases (38.5%), the larger individuals behave as aggressors; in 5 other cases, aggressors were of the same size class of the subordinates; and 3 times (23.1%) the smaller individuals were the aggressors. The aggressors never left, and either showed weak reactions (regardless of the size of the fish attacked) in 3 cases or remained near the bait (10). The subordinates in 2 cases escaped and did not return (in both they were of the same size class of the aggressor), but in most cases they showed weak reactions and remained near the bait (Video S1).

Figure 5. Sequence of an agonistic event. Two toothfish are involved (left side of the images) on the side of a third individual which is eating the bait. Video recorded at Site 06, Terra Nova Bay.

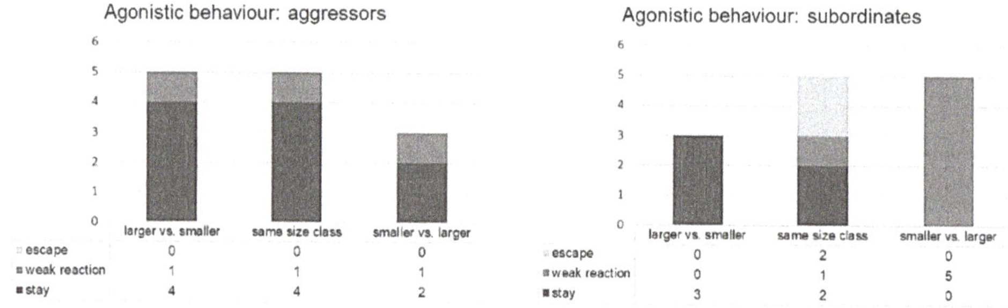

Figure 6. Reaction of aggressors and subordinates in the 13 recorded agonistic events, according to the size of the involved individuals.

3.2. BRUV Data and Longline Catch

Sampling was performed in McMurdo Sound on the shelf area from 537 to 579 m depth (Table 1), at sites characterized by soft substrate bottom (Table S2). While BRUV was deployed for approximately 20 h at each station, in two out of the three cases, the bait was consumed prior to the end of the deployment, about one hour and after about four hours, respectively. Given the key role of the bait in this sampling system, portions of the video recorded in absence of the bait were discarded, and for the sake of standardization, the analysis of all videos was limited to the first hour of setting.

Table 1. McMurdo Sound, total number of fish recorded with BRUV during the first hour of setting at each station, and average vertical longline (VLL) catch values obtained from three deployments at each station.

Station #	Fish Recorded	VLL Catch (Mean ± SD)
20	5	13.0 ± 2.6
28	3	13.7 ± 5.5
29	0	6.3 ± 5.9

During the first hour of the three video samplings, eight Antarctic toothfish were observed. Five (62.5%) were assigned to the immature size class, and three (37.5%) were maturing individuals.

3.3. Sea Bottom Features, Depth, and Toothfish Abundance

A total of 18 Antarctic toothfish were recorded in four out of seven BRUV samplings carried out in 2018 at Terra Nova Bay (Table S2). Most fish were in the maturing size class

(n = 10, 55.6%), but mature (n = 5, 27.7%) and immature fish (n = 3, 16.7%) were also sighted. Eight out of ten maturing toothfish were recorded at a depth higher than 500 m (Site 06).

At sites where no toothfish were recorded, the substrate was mixed and with high percentages of benthos coverage, while in 3 out of 4 sites in which toothfish arrived to the bait, the substrate was soft and the benthos coverage low, as for McMurdo Sound. While toothfish were sighted in correspondence of ridges shallower than 400 m depth, the highest abundances were recorded in a trench habitat, at more than 500 m depth (Table 2).

Table 2. Environmental variables considered to characterize the 7 sites of video sampling performed in 2018 at Silverfish Bay and relative measured abundance metrics. Geomorph = geomorphology; Mean N = Mean Number; TFA = Time of First Arrival.

Site #	Set Duration (hh:mm:ss)	Depth (m)	Substrate	Benthos	Geomorph.	CPUE	MeanN	TFA (min)
01	01:19:38	258	mixed	80%	Ridge	0.75	1	8.2
02	05:38:43	356	soft	40%	Ridge	0.53	0.6	238.8
03	06:01:05	475	mixed	80%	Trench	0	0	-
04	06:29:34	303	mixed	70%	Ridge	0	0	-
05	01:59:25	179	mixed	90%	Ridge	0	0	-
06	06:14:34	543	soft	10%	Trench	1.92	1.4	60.3
07	05:27:00	390	soft	60%	Ridge	0.37	0.4	167.2

The values of CPUE and MeanN at the sites of sighting were strongly correlated ($r = 0.95$, $p < 0.01$), suggesting they may each be indicating abundance even though sample size is low. TFA did not follow the trend of the other two metrics.

3.4. Other Fish Species Recorded by BRUV

While the target species of the work was the Antarctic toothfish, other fish species, representative of the shelf demersal and pelagic assemblages, were recorded by our mini-BRUV. Juveniles and a large shoal of adult Antarctic silverfish (*Pleuragramma antarctica*) were recorded at McMurdo Sound. Juveniles Antarctic silverfish were also often recorded during the mini-BRUV deployments at Silverfish Bay. In this latter area, various *Trematomus* species were recorded. *Trematomus hansoni* were often seen approaching the bait, occasionally trying to eat it, or just swimming in the camera field in groups of five or more individuals. *Trematomus bernacchii* were also recorded around the bait, usually solitary swimming. One Artedidraconidae and one *Chionodraco* sp. incidentally entered the field of view of the camera; however, they did not show any interest for the bait. A group of *Trematomus borchgrevinki* was recorded in the upper water column, close to the sea-ice during the deployment and the hauling of the BRUV.

4. Discussion

The Antarctic toothfish is a high trophic level predator in the Antarctic ecosystems, utilizing a broad range of habitats during its lifecycle, from the epipelagic realm to bentho-pelagic slope habitats down to 2000 m depth [4]. Current understanding on the biology and life cycle of the species mainly relies on fishery-dependent data [2,3], resulting in catchability biases and knowledge gaps. New data on the abundance and distribution of the Antarctic toothfish in coastal shelf areas are required to support population hypotheses and management of this living resource. Large stretches of the coastal shelf areas are data-poor or unexplored, due to the occurrence of fast sea ice that prevents the access of fishing vessels. This holds truth even in regions where the toothfish is historically harvested, such as the Ross Sea region.

We investigated the performance of a BRUV system to collect biological data on the Antarctic toothfish in the data-poor sea-ice covered areas of the Ross Sea region shelf, a region of interest for fishery management and marine conservation.

BRUV systems were demonstrated to produce results comparable to some fishery-based methods for monitoring trends in the relative abundance [31] and have been proven effective in surveying mobile predators and/or opportunistic scavengers, such as sharks [22,37] and grenadier [49], thus making this sampling technique promising to collect fishery independent data on the predator and scavenger Antarctic toothfish [50]. Prior to the present work, a few attempts were made for the use of BRUV systems in the Antarctic waters. In the West Antarctic Peninsula, a BRUV was used to examine scavenging fauna in relation to the exposure to iceberg scouring within Ryder Bay [26], and a preliminary study was performed on the use of baited cameras mounted on a rather large autonomous lander (Aberdeen University Deep Ocean Submersible (AUDOS)) to estimate the abundance and size of the co-generic Patagonian toothfish *D. eleginoides* in sub-Antarctic areas [24]. In both cases, the activities were in open waters, with rather voluminous baited camera systems deployed from boats and set on the sea bottom.

The study on *D. eleginoides*, while demonstrating the general feasibility of toothfish abundance estimation by BRUV, stressed the need for improvements in the design of the system and suggested the use of short-term deployment times (2–3 h). Following those recommendations, in the present study, the efforts were put in the design and configuration of the system, as well as in the optimization of the protocol.

In line with recent trends towards the development of miniaturized deep-sea cameras targeting the reduction of size and costs [51], our BRUV system was developed to be essential in the design and small in size and set up with the camera view facing vertically downward. The majority of BRUVS set-ups used a horizontal camera arrangement, and only the 14% had a vertical orientation pointing down to the seafloor [21]. The vertical camera setting is underused because some species seem reluctant in entering the vertical field of view, most likely due to the perceived confined space under the camera, emphasized by the occurrence of large aluminium frames. Our essential BRUV system does not present any frames or bait arms (the bait is set close to the weight connected by a short piece of fishing line), while camera and light are suspended at about 2 m from the seafloor. Such a setting did not seem to affect the fish behaviour, and Antarctic toothfish were recorded swimming around the bait for quite long time, about 8 min on average, and up to 44 min. In this configuration, the mini-BRUV system was proven suitable for deployment through holes in the sea ice of relatively small diameter (even less than 40 cm), significantly reducing the workload and logistics required to perform surveys from the sea ice. Furthermore, it was easily transportable and light enough to be set and hauled by hand.

In order to optimize the protocol, soak time was carefully considered. During the field work at McMurdo Sound, the association of BRUV work to the longline activities imposed long duration of deployments, with soak times of over 20 h. However, the a posteriori analysis of the videos clearly demonstrated that long deployments are unnecessary and that soak time from one to six hours is ideal, allowing both data collection and multiple deployments in a short period. Remarkably, no decay in the effect of the bait was recorded in such a timeframe, and Antarctic toothfish were spotted approaching the bait until hour six. Overall, the effectiveness of short-term deployments to monitor *Dissostichus* species [24] is confirmed; one-hour video record is likely sufficient to collect baseline data; however, deployments up to six hours might allow for a more precise calculation of relative abundance metrics.

A critical point for the optimization of protocols for BRUV observations is the effectiveness of the bait in attracting the target species, which has repercussions on abundance metrics [39,52,53]. Here, the bait was a squid routinely used in the Antarctic toothfish longline fishery. The behaviour of the fish at the bait supports a positive olfactory response and searching behaviour in the odour plume, which is expected based on their olfactory capabilities [16]. The fish were observed approaching slowly to the bait, often sliding it along their flank and eventually grasping it in their jaws. *Dissostichus eleginoides* individuals were reported to be attracted by the bait but never observed to investigate closely or attempt to take the bait [24]. Such distinct behaviours could be related to differences in the BRUV configurations. In particular, the essential design of our mini-BRUV was likely perceived by fish as not disturbing and did not arise suspicion on fish. However, it is also worth noting that the work on *D. eleginoides* was performed in a commercially longlined, plenty-of-food area, while our activity on *D. mawsoni* was performed in areas not accessible to fishing vessels and, in the case of Terra Nova Bay, in the frame of a specially protected area (ASPA 173) where disturbance of anthropic origin was minimal.

The effectiveness of the bait in attracting Antarctic toothfish individuals was also supported by the absence of any fish record in the negative controls held with no bait. The effect of other potential attractants related to the BRUV, such as light and noise, or presence of other organisms around the system, seems negligible. Interestingly, while the use of intense light in BRUV is usually discouraged due to possible flash induced bait shyness [24], the Antarctic toothfish individuals observed under the light of our mini-BRUV did not show any discomfort; on the contrary, some of them were attracted by the light and occasionally were recorded swimming upward pointing the lamp (Video S1).

The design and configuration of our mini-BRUV allowed to collect high-quality video imagery of 60 Antarctic toothfish in 13 deployments. To the best of our knowledge, only brief images of a single Antarctic toothfish, incidentally acquired during a video survey in McMurdo Sound [54], and sparse snapshots of fish from cameras attached to Weddell seals [55] were available prior to this work. Here, the high quality of the videos is coupled with high number of records, thus providing valuable new documentation of the Antarctic toothfish in its natural habitat (Video S1). Owing to the quality of the videos and the low number of individuals occurring in the field of view simultaneously, identification was possible for all the individuals. Scars and unique colour patterns present were effective natural markings to distinguish individuals and, although the persistence over time of those markings is unknown, at least they can be used as effective markers within a single deployment. This facilitated the abundance metric calculations, and biases related to re-counting were avoided.

Two of the three used metrics, CPUE and MeanN, resulted strongly correlated, but their efficacy remains to be confirmed with additional data. The possibility to use MeanN as a proper abundance metric would allow to avoid the identification of single individuals, a difficult and time-consuming step necessary for the calculation of CPUE, thus significantly reducing the time allocated to video processing, which is necessary. This would foster the application of the methodology and the collection of a large number of recordings. Among the other metrics considered, the time of first arrival (TFA) was successfully adopted for the congeneric Patagonian toothfish [24]. Such a metric is powerful when integrated with data on the current and fish swimming speed [51]. However, in the present study, due to the lack of current speed measurements and reliable estimate of the speed of fish attracted by the bait, the TFA could not be standardized, and its informative value as proxy of Antarctic toothfish abundance was poor.

The low number of BRUV deployments for comparison with longline catch in McMurdo Sound prevented any robust calculation of metrics or trends. However, at the stations where the catch with longlines was the highest, Antarctic toothfish individuals appeared in the field of view of the camera during the first hour, supporting the relation between fish catches and BRUV counts. A relation between the calculated BRUV abundance metrics and the longline catch rates seems to be occurring, but further comparative investigations are needed to provide statistical support to this observation.

Besides quantifying the relative abundance and distribution of target species, the BRUV systems hold potential to generate a variety of data; characterize benthic habitats; and assess functional diversity, body sizes, and animal behaviours [22], thus, in turn, facilitating investigation of fish–habitat relationship [56]. We collected basic information on the sea bottom features (composition and granulometry) from all videos and conducted a preliminary study to investigate the relationship between physical characteristics of the environment and presence of Antarctic toothfish at Terra Nova Bay. Regardless of depth and geomorphology, toothfish were recorded at three stations characterized by soft bottom and relatively low benthos coverage and only one station with a mixed sea bottom. The preference toward soft sea bottom seems confirmed by the data from McMurdo Sound, where toothfish were BRUV sampled only in soft bottom areas.

Another interesting observation from the videos is the size of the Antarctic toothfish at Terra Nova Bay, larger than that expected for the Ross Sea shelf area [3], with prevalence of individuals between 100 and 130 cm length (55.6%) and some larger mature ones (27.7%). It is worth noting that the fish sampled by BRUV at McMurdo Sound are overall smaller than the ones observed at Terra Nova Bay. Despite being a point observation, this aspect might have repercussions on current population hypothesis for the Ross Sea and deserves further investigation.

Overall, the results of our trials, encourage the use of BRUV to study the abundance and distribution of Antarctic toothfish in sea-ice covered areas but also as a valid investigation tool for field work in other areas around the Antarctic continent, whether or not seasonally covered with fast ice. Here, clues for the optimization of the sampling protocols, including information on the bait and optimal soak time, are provided. Furthermore, this study, although preliminary, identified gaps in the knowledge base that could be addressed by the use of BRUV, including habitat preferences of the Antarctic toothfish and size distribution in ice-covered shelf areas. As a side information, our mini-BRUV was proven effective in attracting fish species other than our target species and could thus be used to study densities and behaviour of those species, as well as fish assemblages.

Our pilot investigation allowed to highlight that some improvements should be considered in future studies. In particular, the addition of a current meter would allow to consider current speed that might influence the odour dispersal and lead to bias in the metrics calculations, especially TFA [24]. Another important improvement would be the addition of coupled lasers pointer with parallel light beams, which would allow a more accurate calculation of the fish length and evaluation of the size distribution for the species. Furthermore, the laser pointers would aid in estimating the fish swimming speed, another key element that influences the TFA and, consequently, the abundance estimate. The setting up on the BRUV of sensors for water parameters might add relevant environmental information in support of habitat preferences evaluations. The description of water mass preferences of adult toothfish might provide relevant information on the movements of adults on the continental shelf, with repercussion on the management of the species. The use of non-extractive methods, including BRUV, is of particular relevance in the Ross Sea region, where the largest MPA of the world has recently been established and where this tool can be used in many more applications to study target species but also assemblages and behaviour, supporting research and monitoring in the area.

Supplementary Materials: The following are available online at https://www.mdpi.com/2077-1312/9/3/255/s1. Video S1: Behaviour of Antarctic toothfish (video recorded on November 14, 2018, at Site 06, Terra Nova Bay). Table S2: Summary table.

Author Contributions: Conceptualization, L.G. and M.V.; methodology, D.D.B. and E.C.; validation, E.P.; formal analysis, D.D.B.; investigation, D.D.B., E.C., S.C., and S.J.P.; resources, S.C.; data curation, D.D.B.; writing—original draft preparation, D.D.B.; writing—review and editing, E.P., L.G., and S.J.P.; supervision, L.G., S.J.P., and M.V.; project administration, L.G.; funding acquisition, L.G., M.V., and S.J.P. All authors have read and agreed to the published version of the manuscript.

Funding: This research was funded by the Italian National Programme for Antarctic Research (projects MIUR-PNRA 2015/B1.02 – DISMAS, MIUR-PNRA 2016/AZ1.19 – PILOT) and by the New Zealand Ministry for Primary Industries (project ANT201501) and Antarctica New Zealand.

Institutional Review Board Statement: In situ activities were carried out at McMurdo Sound in accordance with permit AMLR15/R01/Parker/K086 issued by the New Zealand government under the Antarctic Marine Living Resources (AMLR) Act 1981 and at Terra Nova Bay in compliance with the Protocol of Environmental Protection to the Antarctic Treaty, Annex II, Art.3, in the frame of PNRA Research Projects.

Data Availability Statement: The data presented in this study are available in article and in the Supplementary Materials.

Acknowledgments: We thank the CCAMLR EMM and FSA Working Groups participants, with whom the work was discussed, for fruitful suggestions on the refinement of the technique. The participation of Davide Di Blasi in the CCAMLR Working Groups was allowed by the CCAMLR Scholarship for early career researchers during years 2018–2019. Part of the work was carried out during his PhD at the University of Genoa. We thank the three anonymous reviewers for their thoughtful comments and constructive remarks.

Conflicts of Interest: The authors declare no conflict of interest. The funders had no role in the design of the study; in the collection, analyses, or interpretation of data; in the writing of the manuscript; or in the decision to publish the results.

References

1. Long, R.D.; Charles, A.; Stephenson, R.L. Key principles of marine ecosystem-based management. *Mar. Policy* **2015**, *57*, 53–60. [CrossRef]
2. Mormede, S.; Dunn, A.; Hanchet, S.M. A stock assessment model of Antarctic toothfish (*Dissostichus mawsoni*) in the Ross Sea region incorporating multi-year mark-recapture data. *CCAMLR Sci.* **2014**, *21*, 39–62.
3. Hanchet, S.M.; Dunn, A.; Parker, S.J.; Horn, P.L.; Stevens, D.W.; Mormede, S. The Antarctic toothfish (*Dissostichus mawsoni*): Biology, ecology, and life history in the Ross Sea region. *Hydrobiologia* **2015**, *761*, 397–414. [CrossRef]
4. Hanchet, S.M.; Rickard, G.J.; Fenaughty, J.M.; Dunn, A.; Williams, M.J. A hypothetical life cycle for Antarctic toothfish *Dissostichus mawsoni* in Antarctic waters of CCAMLR Statistical Area 88. *CCAMLR Sci.* **2008**, *15*, 35–54.
5. Parker, S.J.; Stevens, D.W.; Ghigliotti, L.; La Mesa, M.; Di Blasi, D.; Vacchi, M. Winter spawning of Antarctic toothfish *Dissostichus mawsoni* in the Ross Sea region. *Antarctic Sci.* **2019**, *31*, 243–253. [CrossRef]
6. Eastman, J.T.; DeVries, A.L. Aspects of body size and gonadal histology in the Antarctic toothfish, *Dissostichus mawsoni*, from McMurdo Sound, Antarctica. *Polar Biol.* **2000**, *23*, 189–195. [CrossRef]
7. Ramorino, M.C. Rapporto sulla Campagna Antartica Estate Australe 2003–2004; Diciannovesima spedizione. In *PNRA Programma Nazionale di Ricerche in Antartide*; Progetto Antartide, Final Report; Consortium for the Implementation of National Research Programmes in Antarctica: Roma, Italy, 2004; pp. 71–75.
8. Ainley, D.G.; Nur, N.; Eastman, J.T.; Ballard, G.; Parkinson, C.L.; Evans, C.W.; DeVries, A.L. Decadal trends in abundance, size and condition of Antarctic toothfish in McMurdo Sound, Antarctica, 1972–2011. *Fish Fish* **2013**, *14*, 343–363. [CrossRef]
9. Parker, S.J.; Mormede, S.; DeVries, A.; Hanchet, S.M.; Eisert, R. Have Antarctic toothfish returned to McMurdo Sound? *Antarctic Sci.* **2016**, *28*, 29–34. [CrossRef]
10. Parker, S.J.; Mormede, S.; Hanchet, S.M.; DeVries, A.; Canese, S.; Ghigliotti, L. Monitoring Antarctic toothfish in McMurdo Sound to evaluate the Ross Sea region Marine Protected Area. *Antarctic Sci.* **2019**, *31*, 195–207. [CrossRef]
11. Mallet, D.; Pelletier, D. Underwater video techniques for observing coastal marine biodiversity: A review of sixty years of publications (1952–2012). *Fish. Res.* **2014**, *154*, 44–62. [CrossRef]
12. Ono, K.; Kotwicki, S.; Dingsør, G.E.; Johnsen, E. Multispecies acoustic dead-zone correction and bias ratio estimates between acoustic and bottom-trawl data. *ICES J. Mar. Sci.* **2018**, *75*, 361–373. [CrossRef]
13. O'Driscoll, R.L.; Ladroit, Y.; Parker, S.J.; Vacchi, M.; Canese, S.; Ghigliotti, L.; Dunford, A.J.; Mormede, S. Acoustic deployments reveal Antarctic silverfish under ice in the Ross Sea. *Antarctic Sci.* **2018**, *30*, 345–353. [CrossRef]
14. Cappo, M.; Harvey, E.; Shortis, M. Counting and Measuring Fish with Baited Video Techniques-An Overview. In *Australian Society for Fish Biology Workshop Proceedings*; Australian Society for Fish Biology: Tasmania, Australian, 2006; Volume 1, pp. 101–114.
15. Bassett, D.K.; Montgomery, J.C. Investigating nocturnal fish populations in situ using baited underwater video: With special reference to their olfactory capabilities. *J. Exp. Mar. Biol. Ecol.* **2011**, *409*, 194–199. [CrossRef]
16. Ferrando, S.; Amaroli, A.; Gallus, L.; Di Blasi, D.; Carlig, E.; Rottigni, M.; Vacchi, M.; Parker, S.J.; Ghigliotti, L. Olfaction in the Antarctic toothfish *Dissostichus mawsoni*: Clues from the morphology and histology of the olfactory rosette and bulb. *Polar Biol.* **2019**, *42*, 1081–1091. [CrossRef]
17. Carlig, E.; Di Blasi, D.; Ghigliotti, L.; Pisano, E.; Faimali, M.; O'Driscoll, R.; Parker, S.; Vacchi, M. Diversification of feeding structures in three adult Antarctic nototheniid fish. *Polar Biol.* **2018**, *41*, 1707–1715. [CrossRef]
18. Bailey, D.M.; King, N.J.; Priede, I.G. Cameras and carcasses: Historical and current methods for using artificial food fall to study deepwater animals. *Mar. Ecol. Prog. Ser.* **2007**, *350*, 179–191. [CrossRef]
19. Dunlop, K.M.; Marian Scott, E.; Parsons, D.; Bailey, D.M. Do agonistic behaviours bias baited remote underwater video surveys of fish? *Mar. Ecol.* **2015**, *36*, 810–818. [CrossRef]
20. Ellis, D.M.; DeMartini, E.E. Evaluation of a video camera technique for indexing the abundance of juvenile pink snapper, *Pristipomoides filamentosus*, and other Hawaiian insular shelf fishes. *Fish B-NOAA* **1995**, *93*, 67–77.
21. Whitmarsh, S.K.; Fairweather, P.G.; Huveneers, C. What is Big BRUVver up to? Methods and uses of baited underwater video. *Rev. Fish Biol. Fisher.* **2017**, *27*, 53–73. [CrossRef]

22. Devine, B.M.; Wheeland, L.J.; Fisher, J.A. First estimates of Greenland shark (*Somniosus microcephalus*) local abundances in Arctic waters. *Sci. Rep.* **2018**, *8*, 1–10. [CrossRef] [PubMed]
23. Devine, B.M.; Wheeland, L.J.; de Moura Neves, B.; Fisher, J.A. Baited remote underwater video estimates of benthic fish and invertebrate diversity within the eastern Canadian Arctic. *Polar Biol.* **2019**, *42*, 1323–1341. [CrossRef]
24. Yau, C.; Collins, M.A.; Bagely, P.M.; Everson, I.; Nolan, C.P.; Priede, I.G. Estimating the abundance of Patagonian toothfish *Dissostichus eleginoides* using baited cameras: A preliminary study. *Fish. Res.* **2001**, *51*, 403–408. [CrossRef]
25. Smale, D.A.; Barnes, D.K.A.; Fraser, K.P.P.; Mann, P.J.; Brown, M.P. Scavenging in Antarctica: Intense variation between sites and seasons in shallow benthic necrophagy. *J. Exp. Mar. Biol. Ecol.* **2007**, *349*, 405–417. [CrossRef]
26. Dunlop, K.M.; Barnes, D.K.A.; Bailey, D.M. Variation of scavenger richness and abundance between sites of high and low iceberg scour frequency in Rider Bay, west Antarcic Peninsula. *Polar Biol.* **2014**, *37*, 1741–1754. [CrossRef]
27. Vanney, J.R.; Falconer, R.K.H.; Johnson, G.L. Geomorphology of the Ross Sea and adjacent oceanic provinces. *Mar. Geol.* **1981**, *41*, 73–102. [CrossRef]
28. Langhorne, P.J.; Hughes, K.G.; Gough, A.J.; Smith, I.J.; Williams, M.J.M.; Robinson, N.J.; Stevens, C.L.; Rack, W.; Price, D.; Leonard, G.H.; et al. Observed platelet ice distributions in Antarctic sea ice: An index for ocean-ice shelf heat flux. *Geophys. Res. Lett.* **2015**, *42*, 5442–5451. [CrossRef]
29. Hoppmann, M.; Richter, M.E.; Smith, I.J.; Jendersie, S.; Langhorne, P.J.; Thomas, D.N.; Dieckmann, G.S. Platelet ice: The Southern Ocean's hidden ice: A review. *Ann. Glaciol.* **2020**, 1–28. [CrossRef]
30. Malandrino, M.; Abollino, O.; Buoso, S.; Casalino, C.E.; Gasparon, M.; Giacomino, A.; La Gioia, C.; Mentasti, E. Geochemical characterisation of Antarctic soils and lacustrine sediments from Terra Nova Bay. *Microchem. J.* **2009**, *92*, 21–31. [CrossRef]
31. Brooks, E.J.; Sloman, K.A.; Sims, D.W.; Danylchuk, A.J. Validating the use of baited remote underwater video surveys for assessing the diversity, distribution and abundance of sharks in the Bahamas. *Endanger. Species Res.* **2011**, *13*, 231–243. [CrossRef]
32. Barord, G.J.; Dooley, F.; Dunstan, A.; Ilano, A.; Keister, K.N.; Neumeister, H.; Preuss, T.; Schoepfer, S.; Ward, P.D. Comparative population assessments of *Nautilus* sp. in the Philippines, Australia, Fiji, and American Samoa using baited remote underwater video systems. *PLoS ONE* **2014**, *9*, e100799. [CrossRef]
33. Stobart, B.; Díaz, D.; Álvarez, F.; Alonso, C.; Mallol, S.; Goñi, R. Performance of baited underwater video: Does it underestimate abundance at high population densities? *PLoS ONE* **2015**, *10*, e0127559. [CrossRef] [PubMed]
34. Langlois, T.J.; Fitzpatrick, B.R.; Fairclough, D.V.; Wakefield, C.B.; Hesp, S.A.; McLean, D.L.; Harvey, E.S.; Meeuwig, J.J. Similarities between line fishing and baited stereo-video estimations of length-frequency: Novel application of kernel density estimates. *PLoS ONE* **2012**, *7*, e45973. [CrossRef]
35. Lowry, M.; Folpp, H.; Gregson, M.; Suthers, I. Comparison of baited remote underwater video (BRUV) and underwater visual census (UVC) for assessment of artificial reefs in estuaries. *J. Exp. Mar. Biol. Ecol.* **2012**, *416*, 243–253. [CrossRef]
36. White, J.; Simpfendorfer, C.A.; Tobin, A.J.; Heupel, M.R. Application of baited remote underwater video surveys to quantify spatial distribution of elasmobranchs at an ecosystem scale. *J. Exp. Mar. Biol. Ecol.* **2013**, *448*, 281–288. [CrossRef]
37. Santana-Garcon, J.; Newman, S.J.; Harvey, E.S. Development and validation of a mid-water baited stereo-video technique for investigating pelagic fish assemblages. *J. Exp. Mar. Biol. Ecol.* **2014**, *452*, 82–90. [CrossRef]
38. Udyawer, V.; Cappo, M.; Simpfendorfer, C.A.; Heupel, M.R.; Lukoschek, V. Distribution of sea snakes in the Great Barrier Reef Marine Park: Observations from 10 yrs of baited remote underwater video station (BRUVS) sampling. *Coral Reefs* **2014**, *33*, 777–791. [CrossRef]
39. Schmid, K.; Reis-Filho, J.A.; Harvey, E.; Giarrizzo, T. Baited remote underwater video as a promising nondestructive tool to assess fish assemblages in clearwater Amazonian rivers: Testing the effect of bait and habitat type. *Hydrobiologia* **2017**, *784*, 93–109. [CrossRef]
40. Willis, T.J.; Millar, R.B.; Babcock, R.C. Protection of exploited fish in temperate regions: High density and biomass of snapper *Pagrus auratus* (Sparidae) in northern New Zealand marine reserves. *J. Appl. Ecol.* **2003**, *40*, 214–227. [CrossRef]
41. Cappo, M.; De'ath, G.; Speare, P. Inter-reef vertebrate communities of the Great Barrier Reef Marine Park determined by baited remote underwater video stations. *Mar. Ecol. Prog. Ser.* **2007**, *350*, 209–221. [CrossRef]
42. Harvey, E.S.; Cappo, M.; Butler, J.J.; Hall, N.; Kendrick, G.A. Bait attraction affects the performance of remote underwater video stations in assessment of demersal fish community structure. *Mar. Ecol. Prog. Ser.* **2007**, *350*, 245–254. [CrossRef]
43. Colton, M.A.; Swearer, S.E. A comparison of two survey methods: Differences between underwater visual census and baited remote underwater video. *Mar. Ecol. Prog. Ser.* **2010**, *400*, 19–36. [CrossRef]
44. Gladstone, W.; Lindfield, S.; Coleman, M.; Kelaher, B. Optimisation of baited remote underwater video sampling designs for estuarine fish assemblages. *J. Exp. Mar. Biol. Ecol.* **2012**, *429*, 28–35. [CrossRef]
45. Campbell, M.D.; Pollack, A.G.; Gledhill, C.T.; Switzer, T.S.; DeVries, D.A. Comparison of relative abundance indices calculated from two methods of generating video count data. *Fish. Res.* **2015**, *170*, 125–133. [CrossRef]
46. Cappo, M.; Harvey, E.; Malcolm, H.; Speare, P. Potential of Video Techniques to Monitor Diversity, Abundance and Size of Fish in Studies of Marine Protected Areas. In *Aquatic Protected Areas. What Works Best and How Do We Know?* Beumer, J.P., Grant, A., Smith, D.C., Eds.; Australian Society for Fish Biology: Saint Lucia, Australia, 2003; Volume 1, pp. 455–464.
47. Cappo, M.; Speare, P.; De'ath, G. Comparison of baited remote underwater video stations (BRUVS) and prawn (shrimp) trawls for assessments of fish biodiversity in inter-reefal areas of the Great Barrier Reef Marine Park. *J. Exp. Mar. Biol. Ecol.* **2004**, *302*, 123–152. [CrossRef]

48. Stoner, A.W.; Laurel, B.J.; Hurst, T.P. Using a baited camera to assess relative abundance of juvenile Pacific cod: Field and laboratory trials. *J. Exp. Mar. Biol. Ecol.* **2008**, *354*, 202–211. [CrossRef]
49. Priede, I.G.; Merrett, N.R. Estimation of abundance of abyssal demersal fishes; a comparison of data from trawls and baited cameras. *J. Fish Biol.* **1996**, *49*, 207–216. [CrossRef]
50. Roberts, J.; Xavier, J.C.; Agnew, D.J. The diet of toothfish species *Dissostichus eleginoides* and *Dissostichus mawsoni* with overlapping distributions. *J. Fish Biol.* **2011**, *79*, 138–154. [CrossRef] [PubMed]
51. Phillips, B.T.; Licht, S.; Haiat, K.S.; Bonney, J.; Allder, J.; Chaloux, N.; Shomberg, R.; Noyes, T.J. DEEPi: A miniaturized, robust, and economical camera and computer system for deep-sea exploration. *Deep. Sea. Res. Pt I* **2019**, *153*, 103136. [CrossRef]
52. Wraith, J.; Lynch, T.; Minchinton, T.; Broad, A.; Davis, A. Bait type affects fish assemblages and feeding guilds observed at baited remote underwater video stations. *Mar. Ecol. Prog. Ser.* **2013**, *477*, 189–199. [CrossRef]
53. Jones, R.E.; Griffin, R.A.; Januchowski-Hartley, S.R.; Unsworth, R.K.F. The influence of bait on remote video observations in shallow-water coastal environments associated with the North-Eastern Atlantic. *PeerJ* **2020**, *8*, e9744. [CrossRef]
54. Eastman, J.T.; Barry, J.P. Underwater video observation of the Antarctic toothfish *Dissostichus mawsoni* (Perciformes: Nototheniidae) in the Ross Sea, Antarctica. *Polar Biol.* **2002**, *25*, 391–395. [CrossRef]
55. Fuiman, L.; Davis, S.; Williams, T. Behavior of midwater fishes under the Antarctic ice: Observations by a predator. *Mar. Biol.* **2002**, *140*, 815–822. [CrossRef]
56. Langlois, T.; Goetze, J.; Bond, T.; Monk, J.; Abesamis, R.A.; Asher, J.; Barrett, N.; Bernard, A.T.F.; Bouchet, P.J.; Birt, M.J.; et al. A field and video annotation guide for baited remote underwater stereo-video surveys of demersal fish assemblages. *Methods Ecol. Evol.* **2020**, *11*, 1401–1409. [CrossRef]

Journal of
Marine Science and Engineering

MDPI

Article

Environmental Influence on the Spatiotemporal Variability of Spawning Grounds in the Western Guangdong Waters, South China Sea

Yao Lu [1,2], Jing Yu [1,*], Zhaojin Lin [1] and Pimao Chen [1]

1 South China Sea Fisheries Research Institute, Chinese Academy of Fishery Sciences, Guangdong Provincial Key Laboratory of Fishery Ecology and Environment, China Scientific Observing and Experimental Station of South China Sea Fishery Resources & Environment, Ministry of Agriculture and Rural Affairs, Guangzhou 510300, China; luyao@scies.org (Y.L.); scslzj@vip.tom.com (Z.L.); chenpm@scsfri.ac.cn (P.C.)

2 South China Institute of Environmental Sciences, Ministry of Ecology and Environment, Guangzhou 510535, China

* Correspondence: yujing@scsfri.ac.cn

Received: 22 July 2020; Accepted: 13 August 2020; Published: 15 August 2020

check for updates

Abstract: Spawning grounds occupy an important position in the supplementary population of fishery resources, especially in Western Guangdong waters (WGWs) in the northern South China Sea (SCS), where fishery resources are being depleted. This study investigated the environmental effects on the spatiotemporal variability of spawning grounds in WGWs, on the basis of generalized additive models (GAMs) and central spawning-ground gravity (CoSGG) by using satellite and in situ observations. Results showed that 57.2% of the total variation in fish-egg density in WGWs was explained. On the basis of stepwise GAMs, the most important factor was sea surface salinity (SSS), with a contribution of 32.1%, followed by sea surface temperature (SST), water depth, month, and chlorophyll a concentration (Chl-a), with contributions of 10.7%, 8.8%, 2.6%, and 2.6%, respectively. Offshore distance had slight influence on the model, explaining approximately 0.4% of the variation in fish-egg density. In summary, fish eggs in WGWs were mainly distributed in the area with SSS of 32.0–34.0 Practical Salinity Unit (PSU), SST of 24–27 °C, and depth of 0–18 m. CoSGG shifted eastwards by 0.38° N and northwards by 0.26° E from April to June. The distribution of spawning grounds in the WGW was affected by the Western Guangdong coastal current (WGCC), cyclonic circulation, the SCS warm current (SCSWC), and changes in the habitat environment (such as SST). Fish in WGWs tend to spawn in areas with a high seabed slope and steep terrain (near the Qiongzhou Strait).

Keywords: fish eggs; environmental factors; spatial factors; generalized additive model; remote sensing

1. Introduction

Western Guangdong waters (WGWs) are located in the northern South China Sea (SCS). This is an important place for fish spawning, feeding, breeding, and migration [1]. Spawning grounds are water areas for the mating, spawning, hatching, and breeding of fish, shrimp, and shellfish. It is an important place for the survival and reproduction of aquatic organisms, and it plays an important role in the supplementation of fishery resources [2]. The early life history of fish has three stages: fish egg, larval, and juvenile. The amount of early supplementation and survival rate affects stock density [3]. As an important stage in early-life history, fish eggs are the most vulnerable and sensitive stage in their lives, and small-scale environmental changes may also have a dramatic impact on their resource–replenishment process [4,5]. Previous studies showed that the habitat environment affects the distribution [6] and density [7,8] of fish eggs, the stock density of fish spawning [7,8], and fish structure [9]. However, the influence of seabed topography and marine environment on the spatiotemporal distribution of spawning grounds is still unclear. Spawning grounds are the basis for the replenishment of fishery resources [10]. Exploring the quantitative relationship between the density and distribution of fish eggs and marine environment factors not only helps to understand the formation mechanism of spawning grounds but, also, reflects variations in local fishery resources.

The relationship between fishery resources and marine environment are complex, nonlinear, and nonadditive [11]. In a quantitative analysis of the relationship between fish-egg density and marine environment, the method choice is important. General additive models (GAMs) can better demonstrate the nonlinear relationship between dependent and multiple independent variables [12], and they are widely used in quantitative analyses of the relationship between fishery resources and environmental factors [1,13–15]. On GAMs, the development status of *Clupea harengus* eggs distributed in western Scottish waters was observed to be related to its vertical spatial distribution, and the development of fish eggs near the bottom was relatively slow [16]. In the Baltic Sea, temperature had no significant effect on the abundance of *Platichthys flesus* eggs [17]. The distribution of *Engraulis encrasicolus* and *Sardinella aurita* in the Mediterranean Sea was related to the interaction between seabed depth and sea surface chlorophyll a (Chl-a) concentration. The near-shore continental shelf area with a high sea surface chlorophyll concentration was more suitable for spawning [18]. In China, on the basis of GAMs, the spatial distribution of dominant fish eggs (*Stolephorus commersonnii* and *Cynoglossus joyneri*) in the Haizhou Bay of the Yellow Sea was related to the underlying temperature [19], and the distribution of *Gadus macrocephalus* eggs in the Yellow Sea was closely related to environmental factors such as bottom-water temperature, quality, and salinity [20]. Satellite remote-sensing technology provides all-weather, large-scale, and high-resolution marine-surface information, and it was successfully applied to marine-fishery research [1,13,14,21]. In this study, satellite remote-sensing data were applied in the analysis of the environmental effects of the spatiotemporal distribution of spawning grounds in WGWs. The early supplementary mechanism of fishery resources in WGWs was explored, providing a reference for the protection of fish habitats in the SCS.

2. Materials and Methods

2.1. Fishery Data

Fish egg data were obtained from spawning-ground surveys from 2014 to 2015 (April–June). Major species of fish eggs identified in this study were *Trichiurus haumela*, Carangidae, *Nemipteras virgatus*, *Sardinella aurita*, and *Anchoviella commersonii*. The research area was at 110–113° E, 19.5–22° N (Figure 1). Fish eggs were sampled by macroplankton nets with a hauling speed of 1.5 n mile/h and then preserved in 5% formaldehyde solution. The fish eggs were identified by morphological characteristics, including shape, size, chorion, yolk, oil globule, and pigmentation [9,10]. In this study, fish-egg data were grouped by 0.25° × 0.25° grid cells. The unit of fish-egg density was ind $(10\ m)^{-3}$.

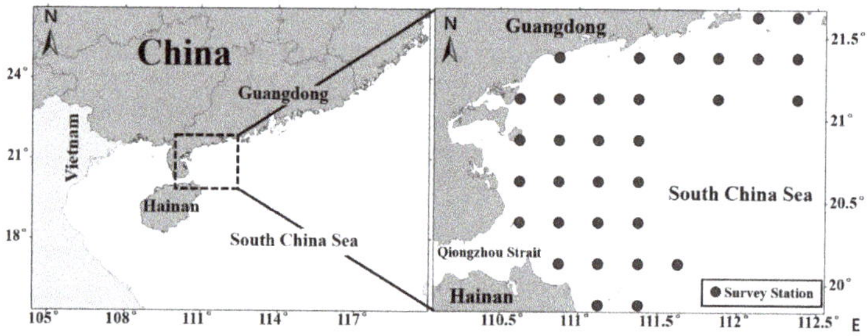

Figure 1. Research area and survey stations of spawning grounds in Western Guangdong waters (WGWs; dotted box, area of satellite data).

2.2. Environmental Data

Satellite data were sea surface temperature (SST), sea surface chlorophyll a concentration (Chl-a), and sea surface salinity (SSS). SST and Chl-a data were derived from MODIS Aqua products of NASA (http://oceancolor.gsfc.nasa.gov), for which temporal resolution was 8 days, and spatial resolution was 4 km. SSS data were obtained from the Global Ocean Physical Reanalysis Product of the Copernicus Marine Environment Management Service (CMEMS, http://marine.copernicus.eu/), for which temporal resolution was one month, and spatial resolution was 1/12°. The digital elevation model (DEM) of WGWs was derived from elevation data of Google Earth, with an elevation level of 18 and a spatial resolution of 8.85 m.

Satellite remote-sensing SST, Chl-a, and SSS data were derived by removing invalid values and performing monthly averaging and data-fusing by using MATLAB (MathWorks, Natick, MA, USA) software. Satellite remote-sensing SST, Chl-a, and SSS data were processed to monthly images through ArcGIS 10.5 (Esri, Redlands, CA, USA)software (ordinary kriging) [22–24]. The distribution of seabed depth (elevation) and seabed-terrain slope was plotted through ArcGIS 10.5 software by DEM data.

2.3. GAMs Fitting Procedures

The GAMs is an additive model that was proposed by Hastie [12]. It is a nonparametric method of generalized linear regression [19]. The primary formulation of this model is

$$Y = \alpha + \sum_{j=1}^{n} f_i(x_j) + \varepsilon \tag{1}$$

where Y, fish-egg density (ind $(10\ m)^{-3}$); x_j, explanatory variable (environmental factors for each survey station); α, formulation intercept; ε, residual; and $f_i(x_j)$, any univariate function of the respective variable with spline smoothing. The formulation of the GAM is

$$\begin{aligned} log(Y+1) = s(Month) + s(Lon) + s(Lat) + s(SST) + s(Chl-a) + s(Slope) + s(SSS) + \\ s(Distance) + s(Depth) + \varepsilon \end{aligned} \tag{2}$$

where Y is the fish-egg density. In order to prevent the response variable from appearing as zero, we made a logarithmic transformation after Y + 1; s(x), spline-smoothing function of covariate x; *Month*, month; *Lat*, latitude; *Lon*, longitude; *SST*, sea surface temperature; *Chl-a*, sea surface chlorophyll a concentration; *Slope*, seabed terrain slope; *SSS*, sea surface salinity; *Distance*, closest distance from the shore; *Depth*, water depth; and ε, model error that obeyed the Gaussian distribution. The mgcv package in R v.3.4.4 software (R Core Team, https://www.r-project.org/) was used to build and test the GAMs [25,26], and a forward-stepwise method was employed to select variables with a significant influence on the model.

The Akaike information criterion (AIC) was applied to check the fitness of the model after adding variables to the model [27]. The smaller the AIC value is, the better the model fit. Generalized cross-validation (GCV) was used to assess predictor variables. The smaller GCV is, the greater the generalization ability of the model [28,29]. The significance and nonlinear contribution of the factor to the nonparametric effect were evaluated by F and chi-squared tests, respectively [30–32].

The formula for calculating the AIC value is

$$AIC = \theta + 2df\varphi \tag{3}$$

where θ, deviation; *df*, effective degree of freedom; and φ, variance.

2.4. Center of Gravity of Spawning Grounds

The center of the spawning-ground gravity (CoSGG) of fish-egg density in each month was calculated with reference to the gravity-center analysis method [1], indicating the spatiotemporal variations of spawning grounds in WGWs. The formula for calculating the CoSGG of spawning grounds is [33]

$$X = \sum_{i=1}^{K}(C_i \times X_i) / \sum_{i=1}^{K} C_i$$

$$Y = \sum_{i=1}^{K}(C_i \times Y_i) / \sum_{i=1}^{K} C_i \quad (4)$$

where X and Y, CoSGG longitude and latitude; C_i, fish-egg density of fishing area i; X_i and Y_i, central latitude and longitude positions of fishing area i; and K, total number of fishing areas.

3. Results

3.1. GAMs Analysis

The spatiotemporal and environmental factors selected on the basis of AIC and GCV values were month (*Month*), sea surface temperature (*SST*), chlorophyll a concentration (*Chl-a*), sea surface salinity (*SSS*), distance from shore (*Distance*), and waters depth (*Depth*). In this study, the GAM formulation was

$$log(Y+1) = s(Month) + s(SSS) + s(Depth) + s(SST) + s(Chl-a) + s(Distance) \quad (5)$$

The deviance explained by this model was 57.2%, with R^2 of 0.531 (Table 1).

Table 1. Deviance analysis for the general additive models (GAMs) fitted to the fish-egg density.

Model Factors	AIC Value	GCV Value	Adjusted R^2 Value	Deviance Explained(%)	Residual Deviance
Log(Y + 1) = NULL	1354.16	6.30	0	0.00	1808.76
Log(Y + 1) = s(Month)	1349.69	6.21	0.02	2.6	1761.76
Log(Y + 1) = s(Month) + s(SSS)	1251.74	4.43	0.32	34.7	1181.80
Log(Y + 1) = s(Month) + s(SSS) + s(Depth)	1214.40	3.89	0.41	43.5	1022.39
Log(Y + 1) = s(Month) + s(SSS) + s(Depth) + s(SST)	1169.28	3.34	0.51	54.2	829.06
Log(Y + 1) = s(Month) + s(SSS) + s(Depth) + s(SST) + s(Chl-a)	1161.02	3.26	0.52	56.8	781.28
Log(Y + 1) = s(Month) + s(SSS) + s(Depth) + s(SST) + s(Chl-a) + s(Distance)	1159.03	3.24	0.53	57.2	774.32

In GAMs, the influence of the spatiotemporal and environmental factors on fish-egg density is indicated by the contributions in Table 2. The most important influencing factor from the selected factors was the SSS with a contribution of 32.1%, followed by SST, depth, month, and Chl-a, with contributions of 10.7%, 8.8%, 2.6%, and 2.6%, respectively. The distance had slight influence on the fish-egg density, explaining 0.4%. As indicated by the ANOVA F-ratio test, all selected factors in the model and fish-egg density showed significant correlations (Pr(F) < 0.05). The chi-squared test evaluates the nonlinear contribution of nonparametric effects, and the lowest value (Pr(chi)) was the best. In the GAMs, the factors of the *SSS*, *SST*, and *Depth* were the best.

AIC, Akaike information criterion; GCV, generalized cross-validation; SSS, sea surface salinity; SST, sea surface temperature; and Chl-a, chlorophyll a concentration.

In GAMs, the influence of spatiotemporal and environmental factors on the fish-egg density were indicated by the contributions in Table 2. The most important influencing factor from the selected factors was the SSS, with a contribution of 32.1%, followed by SST, depth, month, and Chl-a, with contributions of 10.7%, 8.8%, 2.6%, and 2.6%, respectively. Distance had a slight influence on the fish-egg density, explaining 0.4%. As indicated by the ANOVA F-ratio test, all the selected factors in the model and fish-egg density showed significant correlations (Pr(F) < 0.05). The chi-squared test is a type of test to evaluate the nonlinear contribution of nonparametric effects, and the lowest value (Pr(chi)) was the best. In GAMs, the factors of the *SSS*, *SST*, and *Depth* were the best.

Table 2. Contributions of the selected variables in GAMs.

Variables	Contribution (%)	d.f.	Pr(*F*)	Pr(*chi*)
SSS	32.1	10.57	2.2×10^{-16} ***	2.2×10^{-16} ***
SST	10.7	9.02	5.7×10^{-9} ***	4.6×10^{-10} ***
Depth	8.8	2.82	8.3×10^{-9} ***	1.9×10^{-9} ***
Month	2.6	1.81	0.019 *	0.018 *
Chl-a	2.6	4.28	0.004 **	0.004 **
Distance	0.4	0.45	0.047 *	0.046 *

***, $p < 0.001$; **, $p < 0.01$; and *, $p < 0.05$. SSS, sea surface salinity; SST, sea surface temperature; Chl-a, chlorophyll a.; d.f., degrees of freedom; Pr(*F*), *p*-value from an ANOVA F-ration test; and Pr(*chi*), a type of score test to evaluate the nonlinear contributions of the nonparametric effects.

The relationships between the fish-egg density and the chosen factors are presented in Table 2. The environmental factors (SSS, SST, and Chl-a) had the most significant influences on the model, with a total contribution of 45.4% (Table 2). Among those factors, SSS had the greatest influence on the fish-egg density, with a contribution of 32.1% (Table 2). Within the range of 24–30.5 Practical Salinity Unit (PSU), an increase in the SSS had a positive effect on the fish-egg density. As the SSS increased, the confidence interval decreased and reliability increased. Within the range of 30.5–32 PSU, the increase in the SSS had a negative effect on the fish-egg density, with its confidence interval decreasing and credibility increasing. Within the range of 32–34.5 PSU, the SSS increased with the increase in the fish-egg density, reaching its maximum of 34.5 PSU. At the same time, the confidence interval decreased and reliability increased (Figure 2b). The SST contribution to the model was 10.7% (Table 2). Within the ranges of 24–25 and 27–29 °C, the fish-egg density increased with the increasing SST, and it reached its maximum at 29 °C. Within the ranges of 25–27 and 29–32.5 °C, the fish-egg density decreased with the increasing SST, reaching its minimum at 32 °C. Within the range of 29–32.5 °C, the confidence interval increased with the increasing SST, and the reliability was reduced (Figure 2d). Chl-a had a slight influence on the fish-egg density, with a contribution of 2.6% (Table 2). Within the ranges of 0–4 and 10–21 mg m^{-3}, the fish-egg density showed an upward trend, which increased with an increase in the Chl-a, and the fish-egg density reached its maximum at 21 mg m^{-3}. Within the range of 10–21 mg m^{-3}, the confidence interval increased, and the reliability was reduced. Within the ranges of 4–10 mg m^{-3} and 21–35 mg m^{-3}, the fish-egg density showed a downward trend. The fish-egg density decreased as the Chl-a increased. The fish-egg density had its minimum at 33 mg m^{-3}; near the minimal value, the confidence interval was large, and the reliability was low (Figure 2e).

Spatial factors (depth and distance) contributed 9.2% to the model (Table 2). Depth contributed 8.8% to the model (Table 2). In the range of 0–18 m, the fish-egg density increased with the increase in depth and reached its maximum at 18 m. Within the range of 18–45 m, the fish-egg density decreased with the increase in depth, reaching its minimum at around 50 m. Within the range of 45–80 m, the fish-egg density increased with the increase in depth, and the confidence interval increased and the reliability decreased (Figure 2c). The contribution rate of distance to the model was 0.4% (Table 2). There was a negative linear correlation between the fish-egg density and distance, and the fish-egg density decreased with the increase in distance. After the distance reached 40 km, the confidence interval increased, and the reliability decreased (Figure 2f).

The contribution of the time factor (month) to the model was 2.6%. The fish-egg density gradually decreased with the increasing month value, reaching its maximum in April, and it remained at a high level in April and May. In June, it dropped significantly, reaching its minimum (Table 2 and Figure 2a).

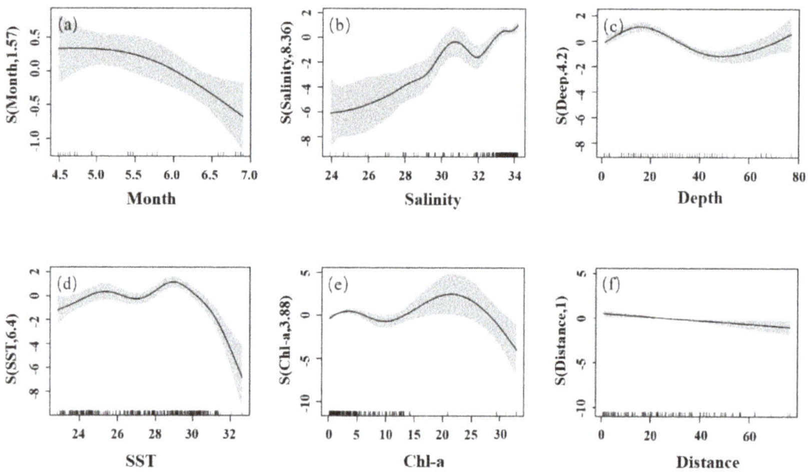

Figure 2. A generalized additive models (GAMs) analysis of the effects of the spatiotemporal and environmental factors on the fish-egg density in WGWs: (**a**) month, (**b**) seas surface salinity (SSS), (**c**) depth, (**d**) sea surface temperature (SST), (**e**) chlorophyll concentration of a (Chl-a), and (**f**) distance. Shadow areas, 95% confidence intervals. Rug plots on the x-axis indicate data density.

3.2. Relationship between Fish-egg Distribution and Environmental Factors

The spatiotemporal distribution of the fish-egg density and SST in WGWs is shown in Figure 3. The SST of the study area was 22–24 °C in April, 24–26 °C in May, and 26–30 °C in June. Areas with a high value of the fish-egg density in April were concentrated in a sea area with the SST of 22–24 °C, in May with the SST of 24–25 °C, and in June with the SST of 28–30 °C (Figure 3a–c). In June, the fish-egg density was generally lower than that in April and May and mainly distributed in high latitudes (20.5–21.5° N) and high-SST (> 28 °C) waters (Figure 3c).

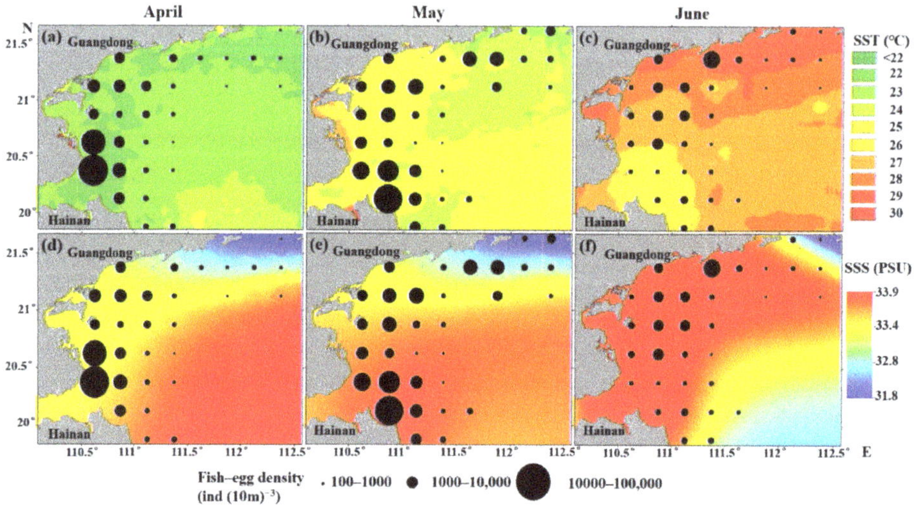

Figure 3. Relationship between the spatiotemporal distributions of the fish-egg density, water temperature, and salinity in WGWs: (**a**) SST in April, (**b**) SST in May, (**c**) SST in June, (**d**) SSS in April, (**e**) SSS in May, and (**f**) SSS in June.

The SSS gradually increased from April to June in WGWs. The high-SSS area tended to move toward the northwestern area (Figure 3d–f). The area with high fish-egg density was concentrated in waters with an SSS of 33.2–33.5 PSU in April, 33.7–33.8 PSU in May, and greater than 33.9 PSU in June.

3.3. Relationship between Fish-egg Distribution and Spatial Factors

The spatial-factor analysis (distance, depth, and slope) and fish-egg density in WGWs showed that the fish-egg density in near-shore areas was generally higher than that in offshore areas (Figure 4a). The fish-egg density was higher in shallow waters (10–20 m) and lower in relatively deep waters (40–70 m) (Figure 4a). In this study, the fish-egg density reached its maximum near the Qiongzhou Strait (32070.05 ind (10 m)$^{-3}$), where the seabed slope was higher (seabed slopes > 1°) than that in other areas (density > 10,000 ind (10 m)$^{-3}$) (Figure 4). The fish-egg density in this area was higher than that in other areas (Figure 4b).

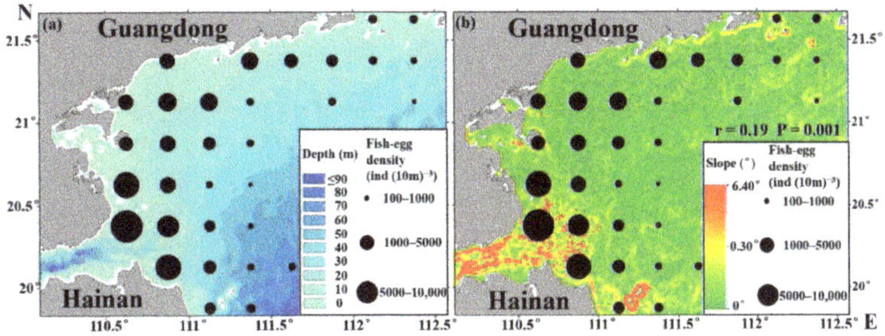

Figure 4. Relationship between the fish-egg density and spatial factors in WGWs: (**a**) distribution of water depth and (**b**) distribution of seabed slope; r, a score of Pearson correlation coefficient; and P, *p*-value from Pearson correlation coefficient.

3.4. CoSGG Variations of Spawning Grounds in WGWs

The CoSGG of the spawning ground in WGW moved from the southwest of the study area to the northeast from April to June (Figure 5a). It was located in the area near Leizhou Peninsula (110.87° E, 20.68° N, green dot in Figure 5a) in April, moved to the eastern waters (111.20° E, 20.72° N, red dot in Figure 5a) in May, and shifted to the northern waters (111.25° E, 20.94° N, yellow dot in Figure 5a) in June. The CoSGG of the spawning grounds varied by 0.03° N and 0.33° E from April to May and 0.22° N and 0.04° E from May to June.

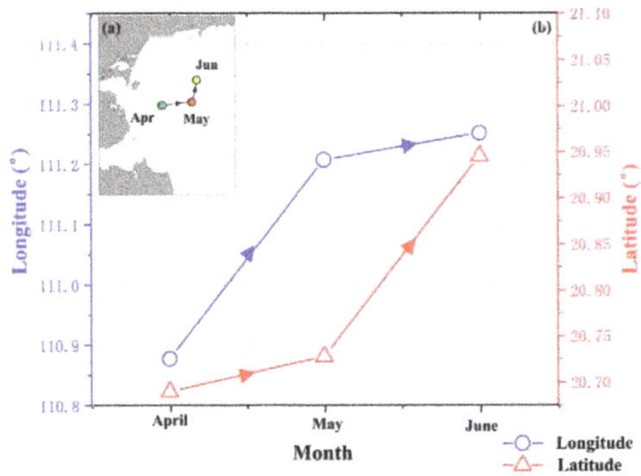

Figure 5. Changes in the central spawning-ground gravity (CoSGG) of the spawning grounds in WGWs: (**a**) moving track and (**b**) longitude and latitude of the CoSGG.

4. Discussion

4.1. Effects of Environmental Factors on Fish-Egg Density

The results of this study highlighted the importance of examining multiple environmental drivers when assessing fish-egg responses to environmental conditions [34,35]. In particular, interactions between geographical and ecological environmental factors obtained from the satellite remote sensing and fish-egg density were found on the basis of GAMs. Among the selected factors, the SSS had the greatest influence on the fish-egg density, with a contribution of 32.1% (Table 2). Fish egg is an early stage of fish-life history. Therefore, the water environment, especially salinity, was one of the major factors affecting the metabolism of fish eggs [36]. Salinity affects embryos developed through changing the osmotic pressure of fish eggs. The research indicated that the hatching rate and increasing salinity showed an approximately normal distribution trend [37]. A higher or lower salinity hinders the material exchange between fertilized eggs and the surrounding medium, resulting in the reduction of the hatching rate and embryo malformation [38]. This study showed that fish eggs in WGWs were mainly distributed in waters with an SSS of 31.5–34.5 PSU, and the most suitable SSS for fish eggs was 33–34.5 PSU (Figure 2b). There were relatively concentrated spawning grounds for *Trichiurus haumela* and *Nemipteras virgatus* in WGWs, where the suitable salinity ranges were 33.0–34.5 [39] and 33.94–34.92 PSU [40], respectively, consistent with the results of this study. In addition, the salinity affected the vertical distribution of fish eggs in the water. In low-salinity waters, fish eggs tended to accumulate, making them unable to get enough oxygen, which was not conducive to the development of the fish eggs. In high-salinity waters, on the other hand, fish eggs could be suspended or floated in water. This could facilitate oxygen absorption from the surrounding water, improving the hatchability of the fish eggs [38,41].

The SSS in the northeastern part of the study area was significantly lower than that in other waters, and the fish-egg density in this area was generally lower than that in other waters (Figure 3d–f) because of the dilution of the Pearl River estuary. From the GAM analysis, the fish-egg density fluctuated with the increase in salinity in the salinity range of 30–34.5 PSU, partly connected with different salinity-suitability levels for various fish-species eggs. The northern SCS is an area with multispecies fishery resources and a complex composition [42]. Suitable salinity (about 34 PSU) provides spawning grounds for fish of different reproductive habits [43]. In addition, the sea-salinity gradient had a certain effect on the spawning grounds. The salinity gradient in Leizhou Bay in China is small, and the salinity has little effect on fish eggs [44]. The salinity gradient in WGWs was larger, and the salinity had a greater impact on the density and spatial distribution of fish eggs.

The contribution rate of the SST to the fish-egg density was 10.7% (Table 2). Fish eggs in WGWs were mainly distributed in waters with SST of 22–32 °C, and the most suitable SST for fish egg survival was 24–30 °C (Figure 2d). Studies showed that the suitable temperature for the fish eggs of *Trichiurus haumela* in the SCS was 25–28 °C [39], consistent with the results of this study. The effects of temperatures at 25 and 29 °C on the fish-egg density had two distinct peak areas (Figure 2d), related to the characteristics of multi-fish-species fishery resources in WGWs [43]. The water temperature is one of the key factors affecting fish metabolism [36,45]. The water temperature affected the number, distribution, and population structure of fish eggs by affecting the adult gonadal development and reproductive migration [46]. The water temperature also had a significant impact on the metamorphosis [6] and hatching speed of fish eggs [47]. The fish-egg density sharply decreased in June, which might have been related to changes in the water temperature. The average water temperature in June was higher (greater than 28 °C; Figure 3c), which reduced the survival rate of some eggs that did not tolerate high temperatures, leading to a decrease in the fish-egg density [48]. On the other hand, an increase in the water temperature promotes the development of some fish eggs, shortens their hatching time, and accelerates the speed of fish-egg hatching, which also leads to a decrease in the fish-egg density [38]. In the East China Sea, the fish-egg density in the summer (June) was higher than that in the spring [49], which was different from the results of this study. This may be related to the difference in the water temperatures in this month. In June, the SST (28–32 °C) in WGWs was higher than the SST (20.03–27.13 °C) in the East China Sea [49]. Different water-temperature levels affected the density and distribution of the fish eggs. The GAM analysis showed that the contribution rate of Chl-a to the fish-egg density was only 2.6% (Table 2). Chlorophyll had little effect on the fish-egg density, because the eggs were in the endotrophic stage and could not prey [6,50].

4.2. Effects of Spatial Factors on Fish-Egg Density

The GAM analysis showed that depth ranked the third in the impact on the fish-egg density, with a contribution of 8.8% (Table 2). The research area was located in the sea with a water depth of 8–55 m. The effect of water depth on the fish-egg density showed a fluctuation trend that first increased, decreased, and then increased (Figure 2c). Water depth was one of the major reasons affecting the spatial distribution of spawning grounds [50]. *Trichiurus haumela* in WGWs mainly spawned in waters with a depth of 40–70 m [39]. *Carangidae* fish spawned in waters with depths lower than 60 m in the spring and summer [33]. *Nemipteras virgatus* mainly inhabited the bottom sediment, with a depth of 60–80 m [40]. The eggs of *Anchoviella commersonii* were mainly distributed in areas with depths of less than 20 m [51]. The eggs of *Sardinella aurita* were mainly distributed in areas with a depth of 10 m [52]. *Anchoviella commersonii* and *Sardinella aurita* of pelagic fish and *Nemipteras virgatus* and *Carangidae* of demersal fish inhabited areas with different water depths. Therefore, the effect of water depth on the fish-egg density in the study area showed a fluctuation trend. In WGWs, the fish-egg density in the deep-water areas was significantly lower than that in shallow–water areas (Figure 4a). The relatively harsh lived-in environments of fish living in deeper waters (less food in shallow waters) and other factors might increase the breeding interval of these fish (annual to perennial). Therefore, the fish-egg density in deep waters was lower than that in other areas [53]. The relevant research found that, in the coastal waters of the North Sea, fish eggs tended to gather in shallow waters (<40 m) and form spawning grounds [50,54]. This was similar to the distribution characteristics of spawning grounds in WGWs. This might be due to the abundance of bait organisms in shallow offshore waters, which can provide an ideal spawning ground for fish [53,55]. Fish eggs cannot swim or move independently [49,56]. Therefore, fish eggs floating in the water are susceptible to the effects of ocean currents. The complex seabed terrain reduced the flow velocity and helped fish eggs to gather in this area. This study showed that there was a significant positive correlation between the fish-egg density and seabed slope ($p < 0.05$; Figure 4b). The Qiongzhou Strait (Figures 1 and 4b) had a high seabed-slope value, and the fish-egg density in this area was also higher than that of other areas (Figure 4). This was because the complex seabed topography provides an ideal environment for fish-spawning communities [19,57]. Previous studies showed that the spawning grounds of the Baltic herring (*Clupea harengus membras*) were also distributed in steep areas with steep slopes on the seabed [58]. Therefore, the seabed slope was one of the conditions for the formation of spawning grounds, and the area with a high seabed-slope value tended to attract fish to spawn.

The distance from shore had the least effect on the fish-egg density, with a contribution of 0.4% (Table 2). The distance of the survey stations in this study was 0-60 km, and the effect on the fish-egg density gradually decreased with the increase in distance (Figure 2f). In this study, the fish-egg density in farther areas was significantly lower than that in offshore areas. This might be connected with the reproductive migration of fish to offshore areas in the spring and the formation of a central spawning ground in offshore areas [49].

4.3. Spatiotemporal Distribution of Fish-Egg Density in WGWs

The center of gravity of the spawning ground in WGWs moved 0.38° E to the east and 0.26° N to the north from the spring to summer (Figure 5). Fish eggs cannot swim [56], and their distribution was related to physical oceanographic factors such as currents and tides [49]. The position of the spawning ground was affected by changes in the ocean currents. Ocean currents such as the Western Guangdong coastal current (WGCC) and the South China Sea warm current (SCSWC) existed in the WGWs (Figure 6) [59]. There was a coastal flow from the Pearl River estuary along the coast of Western Guangdong to the southwest, lastly crossing the Qiongzhou Strait into the Beibu Gulf (Figure 6) [60]. From May to August, a cyclonic circulation was formed in the area centered at 20–20.5° N and 110.75–110° E in WGWs (Figure 6b, c). A cold center was formed in this area, with lower temperatures than those of the surroundings (Figure 3c) [61]. In the outer WGCC area, there is a warm current in the South China Sea (SCSWC) that flows fast northeastward all year round (Figure 6) [2,62]. Therefore, fish eggs were affected by the cyclonic circulation (survey time was in late-May) and drifted eastward to gather northeast of Hainan Island (Figure 6) in April and May. In June, the fish eggs drifted northeastward due to the SCSWC with a fast flow. In WGWs, the fish-egg density in each month showed higher in the west and lower in the east, reaching its maximum in the eastern waters of the Leizhou Peninsula (Figure 6). This was related to the WGCC that flew westward all year round. Fish eggs drifting along the WGCC gathered in the northeastern waters of Hainan Island (Figure 6).

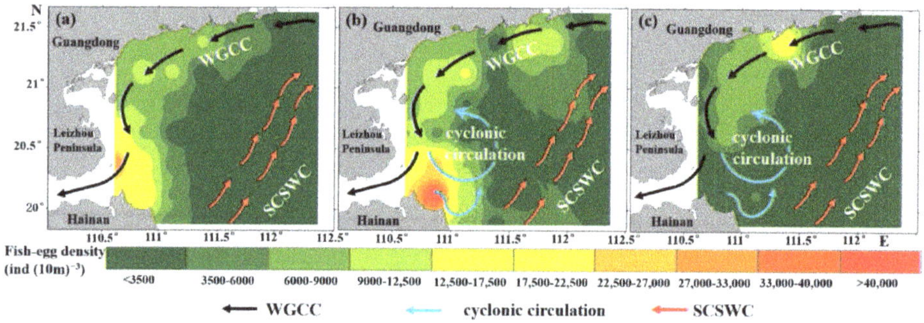

Figure 6. Spatial distribution of the ocean currents and fish eggs in WGWs: (**a**) April, (**b**) May, and (**c**) June (arrow direction represents the current direction). Black arrow, western Guangdong coastal current (WGCC); blue arrow, cyclonic circulation; and red arrow, South China Sea warm current (SCSWC).

5. Conclusions

This study analyzed for the first time the environmental effects of the spatiotemporal distribution of the spawning grounds in WGWs on the basis of satellite remote-sensing and survey data. The most important environmental factor affecting the fish-egg density was the SSS, followed by the SST, depth, month, Chl-a, and distance. The spawning grounds in WGWs were mainly distributed in waters with an SSS of 33.0–34.5 PSU, SST of 24–29 °C, and depth of 5–25 m. The complex seabed terrain was conducive to the accumulation of fish eggs. The results of this study were helpful in understanding the spatiotemporal distribution of early supplementary populations of fishery populations and their response mechanisms to environmental changes in WGWs.

Author Contributions: J.Y. and Y.L. designed the study. Z.L. collected the spawning ground data. Y.L. analyzed the data. P.C. helped the data collection and analysis. J.Y. and Y.L. wrote the article. All authors have read and agreed to the published version of the manuscript.

Funding: This study was supported by the following funds: (1) National Key R&D Program of China (2018YFD0900901), (2) Natural Science Foundation of Guangdong Province, China (2018A030313120), (3) Central Public-interest Scientific Institution Basal Research Fund, CAFS, China (2018HY-ZD0104), (4) R & D Projects in Key Areas of Guangdong Province, China (2020B1111030002), and (5) Special Fund for Basic Scientific Research Business of Central Public Research Institutes (PM-zx703-201904-128).

Conflicts of Interest: The authors declare no conflict of interest.

References

1. Yu, J.; Hu, Q.W.; Tang, D.L.; Chen, P.M. Environmental effects on the spatiotemporal variability of purpleback flying squid in Xisha-Zhongsha waters, South China Sea. *Mar. Ecol. Prog. Ser.* **2019**, *623*, 25–37. [CrossRef]
2. Wang, D.X.; Hong, B.; Gan, J.P.; Xu, H.Z. Numerical investigation on propulsion of the counter-wind current in the northern South China Sea in winter. *Deep Sea Res. Part I Oceanogr. Res. Pap.* **2010**, *57*, 1206–1221. [CrossRef]
3. Hovenkamp, F. Growth-dependent mortality of larval plaice *Pleuronectes platessa* in the North Sea. *Mar. Ecol. Prog. Ser.* **1992**, *82*, 95–101. [CrossRef]
4. Houde, E.D.; Hoyt, R.D. Fish early life dynamics and recruitment variability. In Proceedings of the 10 Annual Larval Fish Conference, Miami, FL, USA, 18–23 May 1986.
5. Peck, M.A.; Huebert, K.B.; Llopiz, J.K. Intrinsic and extrinsic factors driving match–mismatch dynamics during the early life history of marine fishes. In *Advances in Ecological Research*; Elsevier: Amsterdam, The Netherlands, 2012; Volume 47, pp. 177–302.
6. Xiao, Y.Z.; Wang, R.; Qiang, O.; Fang, H.D. Relationship between abundance distribution of fish eggs, larvae and juveniles and environmental factors in the Pearl River Estuary waters in spring. *J. Appl. Oceanogr.* **2010**, *29*, 488–495.
7. Blood, D.M. Low-temperature incubation of walleye pollock (*Theragra chalcogramma*) eggs from the southeast Bering Sea shelf and Shelikof Strait, Gulf of Alaska. *Deep Sea Res. Part II Top. Stud. Oceanogr.* **2002**, *49*, 6095–6108. [CrossRef]
8. Bernal, M.; Somarakis, S.; Witthames, P.R.; Van Damme, C.J.; Uriarte, A.; Lo, N.C.; Dickey-Collas, M. Egg production methods in marine fisheries: An introduction. *Fish Res.* **2012**, *117*, 1–5. [CrossRef]
9. Lin, Z.J.; Wang, X.H.; Jiang, Y.E. Distribution and species composition of fish eggs in Daya Bay. *J. Fish. Sci. China* **2010**, *17*, 543–550.
10. Wan, R.; Zhou, F.; Shan, X.; Sun, S. Impacts of variability of habitat factors on species composition of ichthyoplankton and distribution of fish spawning ground in the Changjiang River estuary and its adjacent waters. *Acta Ecol. Sin.* **2010**, *30*, 155–165. [CrossRef]
11. Stenseth, N.C.; Mysterud, A.; Ottersen, G.; Hurrell, J.W.; Chan, K.; Lima, M. Ecological effects of climate fluctuations. *Science* **2002**, *297*, 1292–1296. [CrossRef]
12. Hastie, T.; Tibshirani, R. *Generalized Additive Models*; Chapman and Hall: London, UK, 1990; pp. 587–602.
13. Yu, J.; Hu, Q.W.; Tang, D.L.; Zhao, H.; Chen, P.M. Response of *Sthenoteuthis oualaniensis* to marine environmental changes in the north-central South China Sea based on satellite and in situ observations. *PLoS ONE* **2019**, *14*, e211474. [CrossRef]
14. Yu, J.; Liu, Z.N.; Chen, P.M.; Yao, L.J. Environmental factors affecting the spatiotemporal distribution of *Decapterus maruadsi* in the western Guangdong waters, China. *Appl. Ecol. Environ. Res.* **2019**, *17*. [CrossRef]
15. Wang, Y.F.; Hu, Q.W.; Yu, J.; Chen, P.M.; Shu, L.M. Effect assessment of fishery resources proliferation in Zhelin Bay marine ranching in eastern Guangdong. *South China Fish. Sci.* **2019**, *15*, 12–19.
16. Stratoudakis, Y.; Gallego, A.; Morrison, J.A. Spatial distribution of developmental egg ages within a herring *Clupea harengus* spawning ground. *Mar. Ecol. Prog. Ser.* **1998**, *174*, 27–32. [CrossRef]
17. Ustups, D.; Müller-Karulis, B.; Bergstrom, U.; Makarchouk, A.; Sics, I. The influence of environmental conditions on early life stages of flounder (*Platichthys flesus*) in the central Baltic Sea. *J. Sea Res.* **2013**, *75*, 77–84. [CrossRef]
18. Schismenou, E.; Giannoulaki, M.; Valavanis, V.D.; Somarakis, S. Modeling and predicting potential spawning habitat of anchovy (*Engraulis encrasicolus*) and round sardinella (*Sardinella aurita*) based on satellite environmental information. *Hydrobiologia* **2008**, *612*, 201. [CrossRef]
19. Li, Z.G. *Distribution of Main Species of Stow Net in the South Yellow Sea based on GAM and Preliminary Study of Characteristics of Ichthyoplankton Assemblages in Haizhou Bay*; Ocean University of China: Qingdao, China, 2013.
20. Zhang, R.Y.; Bian, X.D.; Shan, X.J.; Jin, X.S.; Guan, L.S. Distribution of the age 0 group Pacific cod (*Gadus macrocephalus*) in the Yellow Sea and its relationship with environmental factors. *J. Fish. China* **2018**, *42*, 870–880. [CrossRef]
21. Yu, J.; Chen, P.M.; Tang, D.L.; Qin, C.X. Ecological effects of artificial reefs in Daya Bay of China observed from satellite and in situ measurements. *Adv. Space Res.* **2015**, *55*, 2315–2324. [CrossRef]

22. Müller, D. Estimation of algae concentration in cloud covered scenes using geostatistical methods. In Proceedings of the ENVISAT Symposium, Montreux, Switzerland, 23–27 April 2007.
23. Yu, J.; Hu, Q.W.; Yuan, H.R.; Chen, P.M. Effect assessment of summer fishing moratorium in Daya Bay based on remote sensing data. *South China Fish. Sci.* **2018**, *14*, 1–9.
24. Wang, Y.F.; Yu, J.; Chen, P.M.; Yu, J.; Liu, Z.N. Relationship between spatial-temporal distribution of light falling-net fishing ground and marine environments. *J. Trop. Oceanogr.* **2019**, *38*, 68–76.
25. Wood, S.N. Stable and efficient multiple smoothing parameter estimation for generalized additive models. *J. Am. Stat. Assoc.* **2004**, *99*, 673–686. [CrossRef]
26. Wood, S.N. Fast stable restricted maximum likelihood and marginal likelihood estimation of semiparametric generalized linear models. *J. R. Stat. Soc. Ser. B (Stat. Methodol.)* **2011**, *73*, 3–36. [CrossRef]
27. Venables, W.N.; Dichmont, C.M. GLMs, GAMs and GLMMs: An overview of theory for applications in fisheries research. *Fish. Res.* **2004**, *69*, 319–337. [CrossRef]
28. Hilborn, R.; Mangel, M. *The Ecological Detective: Confronting Models with Data*; Princeton University Press: Princeton, NJ, USA, 1997.
29. Quinn, T.J.; Deriso, R.B. *Quantitative Fish Dynamics*; Oxford University Press: Oxford, UK, 1999.
30. Stone, C.J. Additive regression and other nonparametric models. *Ann. Stat.* **1985**, *13*, 689–705. [CrossRef]
31. Akaike, H. Information theory and an extension of the maximum likelihood principle. In *Breakthroughs in Statistics*; Springer: New York, NY, USA, 1992; pp. 610–624.
32. Burnham, K.P.; Anderson, D.R. *Model. Selection and Multimodel Inference: A Practical Information-Theoretic Approach*; Springer Science & Business Media: Berlin/Heidelberg, Germany, 2003.
33. Chen, G.B.; Li, Y.Z. Distribution of the Carangidae fishes in the continental shelf waters of northern South China Sea. *J. Shanghai Ocean Univ.* **2003**, *12*, 146–151. [CrossRef]
34. La Mesa, M.; La Mesa, G.; Catalano, B.; Jones, C.D. Spatial distribution pattern and physical–biological interactions in the larval notothenioid fish assemblages from the Bransfield Strait and adjacent waters. *Fish. Oceanogr.* **2016**, *25*, 624–636. [CrossRef]
35. Bacha, M.; Jeyid, M.A.; Vantrepotte, V.; Dessailly, D.; Amara, R. Environmental effects on the spatio-temporal patterns of abundance and distribution of *Sardina pilchardus* and *sardinella* off the Mauritanian coast (North-West Africa). *Fish. Oceanogr.* **2017**, *26*, 282–298. [CrossRef]
36. Singh, A.K.; Hasnain, S.I.; Banerjee, D.K. Grain size and geochemical partitioning of heavy metals in sediments of the Damodar River—A tributary of the lower Ganga, India. *Environ. Geol.* **1999**, *39*, 90–98. [CrossRef]
37. Zhang, J.; Zhang, P.; Chen, Z.Z.; Chen, G.B.; Zhang, K.; Xue, Y.W.; Sun, M.S. Biomass and distribution of carangoid fish resources in the offshore South China Sea. *South China Fish. Sci.* **2016**, *12*, 38–48.
38. Giffard-Mena, I.; Hernández-Montiel, Á.H.; Pérez-Robles, J.; David-True, C. Effects of salinity on survival and plasma osmolarity of *Totoaba macdonaldi* eggs, larvae, and juveniles. *J. Exp. Mar. Biol. Ecol.* **2020**, *526*, 151339. [CrossRef]
39. Lin, J.Q. Studies on the natural regulaive adaptibility of the Hairtail (*Trichiurus haumela*) in the oceanic environment. *Trans. Oceanol. Limnol.* **1981**, *3*, 60–65.
40. Zhang, R.Z.; Lu, S.F. On the eggs and larvae of the golden-thread *Nemipterus Virgatus* (Houttuyn). *Curr. Zool.* **1980**, *26*, 39–42.
41. Nissling, A.; Nyberg, S.; Petereit, C. Egg buoyancy of flounder, *Platichthys flesus*, in the Baltic Sea—Adaptation to salinity and implications for egg survival. *Fish. Res.* **2017**, *191*, 179–189. [CrossRef]
42. Zeng, B.G.; Zhang, S.J.; Chen, G.X. *Investigation and Division of Fishery Resources in the South China Sea*; In Chinese with English Abstract; Guangdong Science and Technology Press: Guangdong, China, 1989.
43. Loots, C.; Vaz, S.; Koubbi, P.; Planque, B.; Coppin, F.; Verin, Y. Inter-annual variability of North Sea plaice spawning habitat. *J. Sea Res.* **2010**, *64*, 427–435. [CrossRef]
44. Yang, Y.Y.; Gao, Y.J.; Wang, J.P.; Xu, B.Q.; Sun, C.X. Community structure of ichthyoplankton and its relationship with environmental factors in Laizhou Bay. *Chin. J. Ecol.* **2018**, *37*, 2976–2984.
45. Edwards, M.; Richardson, A.J. Impact of climate change on marine pelagic phenology and trophic mismatch. *Nature* **2004**, *430*, 881. [CrossRef] [PubMed]
46. Chen, X.Q.; Ruan, C.X.; Yuan, C.G. Effects of different temperature and food on growth and survival rates of the larvae of *Trichogaster trichopterus*. *J. Fuji. Fish.* **2008**, *30*, 9–12.

47. Wan, R.J.; Meng, Z.N. The artificial insemination and hatching of *Trichiurus lepturus*. *J. Fish. China* **2003**, *27*, 188–192.
48. Targońska, K.; Żarski, D.; Kupren, K.; Palińska-żarska, K.; Mamcarz, A.; Kujawa, R.; Skrzypczak, A.; Furgała-Selezniow, G.; Czarkowski, T.K.; Hakuć-Błażowska, A.; et al. Influence of temperature during four following spawning seasons on the spawning effectiveness of common bream, *Abramis brama* (L.) under natural and controlled conditions. *J. Therm. Biol.* **2014**, *39*, 17–23. [CrossRef]
49. Wan, R.J.; Zeng, D.Y.; Bian, X.D.; Ni, X.B. Species composition and abundance distribution pattern of ichthyoplankton and their relationship with environmental factors in the East China Sea ecosystem. *J. Fish. China* **2014**, *38*, 1375–1398.
50. Lelièvre, S.; Vaz, S.; Martin, C.S.; Loots, C. Delineating recurrent fish spawning habitats in the North Sea. *J. Sea Res.* **2014**, *91*, 1–14. [CrossRef]
51. Wu, G.Z. The ecological characteristics of distribution of eggs, larvae and juveniles of the *Engr-aulis japonicus* (Temminck & Schlegel) and *Anchoviella commersonii* in the changjiang river estuary. *Oceanol. Limnol. Sin.* **1989**, *20*, 217–229.
52. Zhang, J.B.; Huang, Z.Y. An investigation on fish eggs and larvae in sea area around planning Yangjiang nuclear plant. *J. Trop. Oceanogr.* **2003**, *22*, 78–84.
53. Fernandez-Arcaya, U.; Drazen, J.C.; Murua, H.; Ramirez-Llodra, E.; Bahamon, N.; Recasens, L.; Rotllant, G.; Company, J.B. Bathymetric gradients of fecundity and egg size in fishes: A Mediterranean case study. *Deep Sea Res. Part I Oceanogr. Res. Pap.* **2016**, *116*, 106–117. [CrossRef]
54. Gibson, R.N.; Nash, R.D.; Geffen, A.J.; Van der Veer, H.W. *Flatfishes: Biology and Exploitation*; John Wiley & Sons: Hoboken, NJ, USA, 2014.
55. Aneer, G. Herring (*Clupea harengus* L.) spawning and spawning ground characteristics in the Baltic Sea. *Fish. Res.* **1989**, *8*, 169–195. [CrossRef]
56. Yin, M.C.; Blaxter, J.H.S. Cruising speeds during early development and starvation of marine fish larvae. *Oceanol. Limnol. Sin.* **1989**, *20*, 1–9.
57. Yi, Y.J.; Zhang, S.H.; Wang, Z.Y. The bedform morphology of Chinese sturgeon spawning sites in the Yangtze River. *Int. J. Sediment. Res.* **2013**, *28*, 421–429. [CrossRef]
58. Šaškov, A.; Šiaulys, A.; Bučas, M.; Daunys, D. Baltic herring (*Clupea harengus membras*) spawning grounds on the Lithuanian coast: Current status and shaping factors. *Oceanologia* **2014**, *56*, 789–804. [CrossRef]
59. Xie, L.L.; Cao, R.X.; Shang, Q.T. Progress of Study on Coastal Circulation near the Shore of Western Guangdong. *J. Guangdong Ocean Univ.* **2012**, *32*, 94–98.
60. Wu, B.Y. A study on the circulation in shelf waters west to zhujiang estuary. *J. Appl. Oceanogr.* **1990**, *9*, 14–21.
61. Yang, S.Y.; Bao, X.W.; Chen, C.S.; Chen, F. Analysis on characteristics and mechanism of current system in west coast of Guangdong Province in the summer. *Acta Oceanol. Sin.* **2003**, *25*, 1–8.
62. Guan, B.X. Evidence for a counter-wind current in winter off the southeast coast of China. *Chin. J. Oceanol. Limn.* **1986**, *4*, 319–332. [CrossRef]

Journal of
Marine Science and Engineering

MDPI

Article

Fish Assemblage Structure Comparison between Freshwater and Estuarine Habitats in the Lower Nakdong River, South Korea

Joo Myun Park [1,*], Ralf Riedel [2], Hyun Hee Ju [3] and Hee Chan Choi [4]

1 Dokdo Research Center, East Sea Research Institute, Korea Institute of Ocean Science and Technology, Uljin 36315, Korea
2 S&R Consultancy, Ocean Springs, MS 39564, USA; ralf.riedel@usm.edu
3 Ocean Policy Institute, Korea Institute of Ocean Science and Technology, Busan 49111, Korea; hhju@kiost.ac.kr
4 Fisheries Resources and Environment Division, East Sea Fisheries Research Institute, National Institute of Fisheries Science, Gangneung 25435, Korea; gmlckschl82@korea.kr
* Correspondence: joomyun.park@kiost.ac.kr; Tel.: +82-54-780-5344

check for updates

Received: 6 June 2020; Accepted: 3 July 2020; Published: 5 July 2020

Abstract: Variabilities of biological communities in lower reaches of urban river systems are highly influenced by artificial constructions, alterations of flow regimes and episodic weather events. Impacts of estuary weirs on fish assemblages are particularly distinct because the weirs are disturbed in linking between freshwater and estuarine fish communities, and migration successes for regional fish fauna. This study conducted fish sampling at the lower reaches of the Nakdong River to assess spatio-temporal variations in fish assemblages, and effects of estuary weir on structuring fish assemblage between freshwater and estuary habitats. In total, 20,386 specimens comprising 78 species and 41 families were collected. The numerical dominant fish species were *Tachysurus nitidus* (48.8% in total abundance), *Hemibarbus labeo* (10.7%) and *Chanodichthys erythropterus* (3.6%) in the freshwater region, and *Engraulis japonicus* (10.0%), *Nuchequula nuchalis* (7.7%) and *Clupea pallasii* (5.2%) in the estuarine site. The fish sampled were primarily small species or the juveniles of larger species at the estuary region, while all life stages of fishes were observed at the freshwater habitats. The diversity patterns of fish assemblages varied greatly according to study site and season, with higher trends at estuarine sites during the warm-rainy season. No significant difference in diversity between freshwater and estuarine sites during the cold-dry season were found. Multivariate analyses of fish assemblage showed spatial and seasonal differences of assemblage structures. Higher effects of between-site variability but not within seasonal variability at each site were observed. Variations in assemblage structures were due to different contributions of dominant species in each habitat. Common freshwater species characterized the fish assemblage in the freshwater region, while marine juveniles were significantly associated with the estuarine habitat. The results from the ecological guild analyses showed distinct ecological roles for freshwater and marine species, and overlapping roles for fish sampled at the fishways. The lower reaches of the Nakdong River are an important ecosystem for both freshwater and marine juveniles. Nakdong River estuarine residents and migrant fishes, however, have been negatively affected by the construction of the weir (gravity dam), due to the obstruction to migration from and to freshwater habitats. Conservation and management policies aimed at minimizing anthropogenic influences on estuary ecosystems should focus on evaluating ecological functions of estuary weirs.

Keywords: fish assemblage; lower reaches; Nakdong River; estuary weir; marine juveniles

1. Introduction

Fish assemblages in coastal habitats influenced by rivers consist of a variety of species with diverse life-history strategies and broad functional categories of freshwater, diadromous and estuarine-dependent species. Lower reaches of river systems are characterized by high biological productivity and diversity [1]. Biological communities in these regions are generally influenced by tides, rainfalls, and anthropogenic disturbances [2–4]. In particular, episodic weather events and anthropogenic stressors have significant effects on the structure and functioning of biological communities through declines of biomass, species richness and abundance [5].

Diadromous fish species in temperate Korean rivers include eels (Anguillidae), salmons (Salmonidae) and sweetfishes (Plecoglossidae) [6–8]. Some coastal fishes, including *Coilia nasus* and *Mugil cephalus*, also exhibit occasional upstream movements from coastal waters to lower river reaches [9,10]. Estuary weirs, however, restrict up or downstream dispersal for migratory fishes due to creation of a sharp salinity gradient, controlling spatial distribution in river catchments [11–13]. Fish assemblages in lower river reaches are greatly affected from anthropogenic structures, such as barriers, dams or weirs at both local and catchment scales [12,14,15], contributing to significant discontinuities especially in diadromous fish assemblages [16].

The Nakdong River estuary is an important spawning and nursery ground for many aquatic animals, supporting a productive fishery [17]. In 1987, an estuary weir was constructed to prevent saltwater intrusion into shallow freshwater habitats. The weir has lowered freshwater flows transporting nutrients and sediments to the estuary, threatening the health of estuarine habitats [13,18], such as shifts in population structure of animal communities. Several studies have reported alteration of animal community structures following construction of dykes, weirs, or any other obstruction in estuary environments [19–21]. Kwak and Huh [20] observed that the dominance of fish assemblages in the coastal Nakdong River estuary shifted from demersal to pelagic species after weir construction. The shifts were mostly due to changes in the sediment depositional dynamics caused by the irregular discharge of freshwater, causing changes in circulation patterns within the estuary. Han et al. [16] also documented the loss of species richness following weir construction. Weirs adversely affected diadromous species by blocking their migration routes, but favored nonnative fishes [16,22].

Although many studies have investigated the community structure of estuarine fishes worldwide, and the impacts of estuary weir on fish assemblage structures [15,23–25], such studies are scarce in areas along the southern coast of Korea. Of the few, Hwang et al. [26] have investigated temporal occurrence patterns of fishes in the western Mangyeong Estuary prior to construction of a seawall in 2010, completely enclosing its lower reaches. Park et al. [27] also reported juvenile fish assemblage in the shallow sandy beach of the lower tip of the Nakdong River estuary. Several previous studies have suggested that seasonal variations in fish abundance are primarily the result of species-specific recruitment [28], with environmental factors as strong determinants, especially water temperature and salinity [29–32]. Because lower river reaches play an important role as the nursery habitats for estuarine residents, juveniles of marine species, and temporary habitats for migratory fishes, research evaluating the impacts of weirs within river reaches needs to be conducted for recommendations on restoring connectivity within estuaries where weirs have been constructed and for determining where weirs should be removed.

Anthropogenic effects on freshwater and estuarine fish assemblages are mostly felt at catchment scales. Research, therefore, needs to focus on the impacts at these large scales [33,34]. Thus, this study aimed to examine variations in species composition and abundance at large spatial scales of fishes inhabiting the lower Nakdong River reaches of Korea by evaluating the ecological function of each habitat and the influence of estuary weir on fish assemblage.

2. Materials and Methods

2.1. Study Area and Sampling

The study area comprised of the lower Nakdong River, located in the southeastern part of the Korean Peninsula (Figure 1). The Nakdong River weir was constructed 6 km north from the mouth of the estuary (Figure 1). Three and two stations were established to investigate the spatio-temporal patterns of fish assemblages at the freshwater and estuary regions (two study sites), respectively. Stations extended from the southern end of the Nakdong River to 30 km north, into the freshwater site. Additional samplings were also conducted at the fishways (two pool fish ladder type and a gate type) within weir; pool ladder type (length = 24 m, width = 1.8 m, slope = 2.9°), gate type (length = 50 m, width = 9 m, slope = n.a.). Sampling depths were approximately 5 m at the freshwater site, and between 10 and 15 m at the estuarine site. The mean tidal range at the study area was 1.2 m for spring tides and 0.4 m for neap tides.

Figure 1. Sampling stations from freshwater (A1, A2 and A3) and estuary (B1 and B2) sites investigating the effects of weir on fish assemblages at the lower Nakdong River, South Korea.

Fish samples were collected monthly from June 2010 to May 2011 at the freshwater and estuarine stations, as well as at fishways within the weir. Samplings were grouped on annual mean temperature for further analysis based on seasonality. Two seasons (i.e., cold and warm seasons) were determined based on temperatures around 17 °C. Temperature data were obtained for Busan province from the Korea Weather Data Open Portal (https://data.kma.go.kr/). Because warm and cold seasons in Korea coincide with rainy and dry seasons, respectively, the seasons were classified as cold-dry (November–March) and warm-rainy (May–October) seasons. Sampling was conducted using a 15 m long, 13 m wide, 3.5 cm mesh bottom trawl with a 1 cm liner covering the codend. Towing speed of ca. 1 knot for 40 min, covering an estimated area of was 16,000 m^2, was conducted at each sampling event. A single tow at each station was carried out during the day (between 09:00 and 12:00) simultaneously at the freshwater and estuary regions. Fishway samplings were conducted at the gate and ladder during flood tides. Fish were sampled, the species recorded, and released. Fishway samplings were conducted only during warm-rainy seasons (between May and October) in both 2010 and 2011. In the ladder type fishways, fishes were collected using cone-traps with 20 mm mesh. A set-net (length 10 m, height 4 m, mesh body 50 mm, mesh bag 20 mm) was installed within the gate-type fishway and fish samples were collected after 12 h. All fish samples caught from fishways were checked for species names, tagged adult specimens using various tags, and then released alive for further fish tracking study. Both surface (between 1 m and 2 m) and bottom (between 10 m and 15 m) water temperatures and salinities were monitored monthly at each sampling location using a portable instrument (Thermo Scientific Orion 3-star). Immediately after capture, fish samples from stations were stored in ice and transported to the laboratory for processing. Taxonomic classification and species name were checked by FishBase [8].

2.2. Ecological Guild and Habitat Type of Fishes

Five ecological guilds were identified for this study. Guilds were estuarine residents (ER), freshwater species (FW), marine occasional visitors (MV), marine juvenile (MJ), and diadromous (catadromous, anadromous, or amphidromous) migrants (DM), according to native Korean fishes [7,35]. Fish were further categorized into pelagic and benthic.

2.3. Data Analyses

Full factorial analyses of covariance (ANCOVA) was used to test the effects of site and depth covarying with the seasonal trends of water temperature. A Bonferroni correction for multiple comparison was used to determine post-hoc significance.

The Shannon–Wiener index (H′) was used to estimate community-level diversity [36]. A logarithmic transformation ($\log(x + 1)$) of fish abundance (number of specimens) was performed to correct for heteroschedasticity and to reduce the weight of overly abundant species in analyses. Three independent two-way ANOVA with an orthogonal design were used to analyze the spatial (two sites) and seasonal (two seasons) effects (independent variables) on fish species richness, abundance and diversity (dependent variables). Prior to ANOVA analysis, homogeneity of variance was tested using Levene's test [37]. Further test onto the dependent variables, the ANOVA tests were also conducted using a split file analysis, with the dependent variable of season paired with site and vice versa. The split file analysis on the ANOVA allowed the data output to be separated by factor to allow visualization of case subsets.

Further inferential and descriptive analyses were performed to assess abundance trends with spatial and temporal patterns. Permutation multivariate analyses of variance (PERMANOVA) on log(abundance + 1) based on Bray–Curtis similarity matrices were conducted [38]. Analysis factors for the PERMANOVA were site (two fixed levels: freshwater and estuary), station (nested within site, two random levels), and season (two fixed levels: cold-dry and warm-rainy seasons). Similarity matrices were used in a PERMANOVA to test for factor effects. In cases in which PERMANOVA detected a significant difference at the 0.05 level, posteriori pairwise PERMANOVA comparisons were used to determine which interaction terms differed significantly among variables within each level of factors. PERMANOVA assigns components of variation (COV) of differing magnitudes to the main factors and interaction between combinations of main factors. The larger COV indicates greater influence of a particular factor or interaction term on the structure of the data [39]. The non-metric multidimensional scaling (nMDS) ordination technique was used to visualize factor effects. To assess statistical significance among factor levels, a canonical analysis of principal coordinates (CAPs) was used [39]. Correlation coefficients between each factor and the canonical axis were used as evidence for species contributions to observed differences. Individual species with both correlations higher than 0.4 and total abundance larger than 1% were plotted on CAP axes 1 and 2 for additional visualization of results.

Statistical software used was Systat (Systat version 18, SPSS Inc., Chicago, IL, USA) and PRIMER v7 with the PERMANOVA+ module [39,40]. A 0.05 level for statistical significance was used in analyses.

3. Results

3.1. Environmental Variables

Water temperature and salinity in the study area varied according to seasonal patterns (Figures 2 and 3). Water temperatures at the freshwater site were the lowest in January (1.5 °C) and the highest in September (25.2 °C). In the estuary, water temperature ranged from 7.3 to 25.7 °C at the surface and from 9.1 to 23.8 °C at the bottom, both depths showing similar trends with season. No significant differences of water temperature between stations within each season were detected (ANCOVA; $P > 0.05$). Peak water temperatures were observed in August and September, for both freshwater and estuarine habitats, and minima during January and February (Figure 2). Salinity in the estuarine region varied seasonally, ranging from 5.0 to 34.1‰ at the surface and from 25.1 to 34.3‰ at the bottom. Significantly lower salinity at the surface was observed during the rainy season in the estuarine stations (Figure 3; ANCOVA, $P < 0.05$).

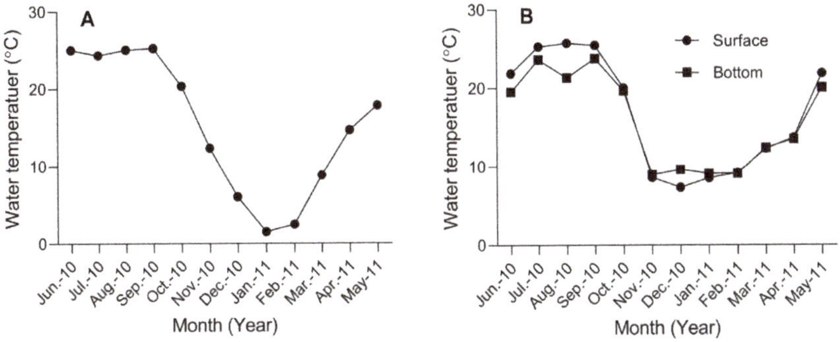

Figure 2. Seasonal changes in the surface and bottom water temperatures at freshwater (**A**) and estuary (**B**) regions at the lower Nakdong River.

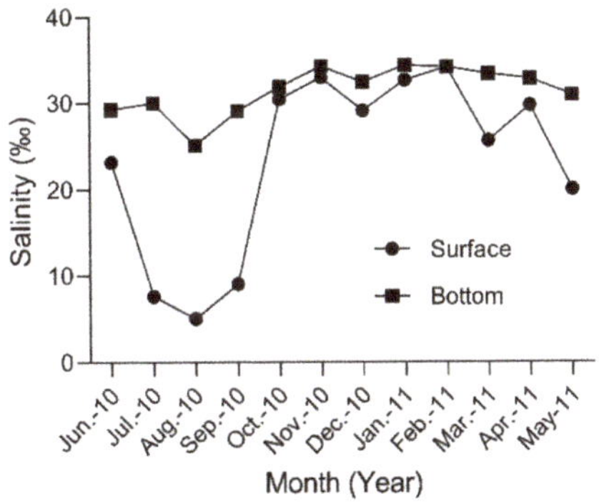

Figure 3. Seasonal changes in the surface and bottom salinities at estuary site at the lower Nakdong River.

3.2. Fish Species Composition

In total, 20,386 individuals belonging to 78 species and 41 families were collected during the study period. Cyprinidae (10 species), Callionymidae (five species), Engraulidae (five species), and Clupeidae (four species) were the most widely represented families (Table 1 and Table S1). Numerically, the five dominant species at the freshwater site were *Tachysurus nitidus*, *Hemibarbus labeo*, *Chanodichthys erythropterus*, *Pseudogobio esocinus* and *Squalidus gracilis*, accounting for 98.0% of the total catch. *Engraulis japonicus*, *Nuchequula nuchalis*, *Clupea pallasii*, *Trachurus japonicus* and *Favonigobius gymnauchen* were the most numerically abundant estuarine fish species, making up 84.7% of the total abundance. Most of the specimens of the dominant species were juveniles of marine fishes, but only *Fa. gymnauchen* was an estuarine resident (Supplementary Material Table S1). In terms of richness, 68 species occurred in the estuarine habitat, which was approximately 3.3–3.5 times more speciouse than freshwater habitats (18 species) or fishways (19 species; Table 1). Only three species co-occurred in both freshwater and estuarine habitats (*Coilia nasus*, *Lateolabrax japonicus* and *Mugil cephalus*), with *Co. nasus* being the most common. In addition, 9 and 12 fishway species co-occurred with fish species in freshwater and estuary habitats, respectively.

Table 1. List of fish species, ecological guild (EG), habitat type (HT), and occurrences of fishes in freshwater (Fr), estuarine habitats (Es), and estuary weir fishway (Fw) stations in the lower Nakdong River, South Korea; ER = estuarine residents, FW = freshwater species, MV = marine occasional visitors, MV(J) = MV included juveniles, MJ = marine juvenile, MD = diadromous migrants (DM); B = benthic, P = pelagic.

Family	Species	EG	HT	Fr	Es	Fw	Family	Species	EG	HT	Fr	Es	Fw
Acropomatidae	*Acropoma japonicum*	MJ	P		•		Hapalogenyidae	*Hapalogenys nigripinnis*	MJ	P		•	
Ammodytidae	*Ammodytes personatus*	MV	P		•		Hemitripteridae	*Hemitripterus villosus*	MJ	B		•	
Apogonidae	*Apogon lineatus*	MJ	P		•		Lateolabracidae	*Lateolabrax japonicus*	MV	P	•	•	•
Bagridae	*Tachysurus nitidus*	FW	B	•		•	Leiognathidae	*Equulites rivulatus*	MJ	P		•	
Callionymidae	*Callionymus beniteguri*	MV	B		•	•		*Nuchequula nuchalis*	MV(J)	P		•	
	Callionymus curvicornis	MV	B		•		Liparidae	*Liparis tanakae*	MV	B		•	
	Callionymus lunatus	MJ	B		•			*Liparis tessellatus*	MJ	B		•	
	Callionymus valenciennei	MV	B		•		Lophiidae	*Lophius litulon*	MJ	B		•	
	Repomucenus virgis	MJ	B		•	•	Monacanthidae	*Rudarius ercodes*	MJ	P		•	
Carangidae	*Trachurus japonicus*	MJ	P		•		Mugilidae	*Mugil cephalus*	DM	P	•	•	•
	Caranx sexfasciatus	MV	P		•		Mullidae	*Upeneus japonicus*	MJ	B		•	
Centrarchidae	*Lepomis macrochirus*	FW	P	•		•	Muraenesocidae	*Muraenesox cinereus*	MV	B		•	
	Micropterus salmoides	FW	P	•			Paralichthyidae	*Paralichthys olivaceus*	MJ	B		•	
Centrolophidae	*Psenopsis anomala*	MJ	P		•			*Tarphops oligolepis*	MJ	B		•	
Clupeidae	*Clupea pallasii*	MJ	P		•		Pholidae	*Pholis fangi*	MV	P		•	
	Clupeidae juvenile sp.	MJ	P		•			*Pholis nebulosa*	MV	P		•	•
	Konosirus punctatus	MJ	P		•		Platycephalidae	*Inegocia japonicus*	MV	B		•	
	Spratelloides gracilis	MJ	P		•			*Platycephalus indicus*	MV	B		•	
Congridae	*Conger myriaster*	MV(J)	B		•		Pleuronectidae	*Kareius bicoloratus*	MV	B		•	•
Cynoglossidae	*Cynoglossus joyneri*	MV	B		•			*Pleuronectes yokohamae*	MV(J)	B		•	•
	Paraplagusia japonica	ER	B		•		Rajidae	*Okamejei acutispina*	MV	B		•	
Cyprinidae	*Acheilognathus macropterus*	FW	P	•				*Okamejei kenojei*	MV	B		•	
	Carassius auratus	FW	P	•			Sciaenidae	*Johnius belengerii*	MJ	P		•	•
	Chanodichthys erythropterus	FW	P	•		•		*Nibea albiflora*	MJ	P		•	
	Cyprinus carpio	FW	P	•				*Pennahia argentata*	MV	P		•	
	Hemibarbus labeo	FW	P	•		•	Scombridae	*Scomber japonicus*	MJ	P		•	
	Hemiculter eigenmanni	FW	P	•			Sillaginidae	*Sillago japonica*	MV	P		•	
	Microphysogobio jeoni	FW	P	•			Siluridae	*Silurus asotus*	FW	B	•		•
	Opsariichthys uncirostris	FW	P	•		•	Sparidae	*Acanthopagrus schlegelii*	MJ	P		•	•
	Pseudogobio esocinus	FW	B	•			Sphyraenidae	*Sphyraena pinguis*	MV	P		•	
	Squalidus gracilis	FW	P	•			Stromateidae	*Pampus argenteus*	MV	P		•	
Engraulidae	*Coilia nasus*	DM	P	•	•		Tetraodontidae	*Lagocephalus wheeleri*	MJ	P		•	
	Engraulis japonicus	MJ	P		•			*Takifugu niphobles*	ER	P		•	
	Thryssa adelae	MV	P		•		Trichiuridae	*Trichiurus lepturus*	MJ	P		•	
	Thryssa hamiltonii	MV	P		•		Triglidae	*Chelidonichthys spinosus*	MJ	B		•	
	Thryssa kammalensis	MV	P		•			*Lepidotrigla guentheri*	MJ	B		•	
Gobiidae	*Acanthogobius flavimanus*	ER	B		•	•		Triglidae sp.	MJ	B		•	
	Favonigobius gymnauchen	ER	B		•	•	Zeidae	*Zeus faber*	MV(J)	P		•	
	Tridentiger brevispinis	ER	B	•		•	Zoarcidae	*Zoarces gillii*	MV(J)	B		•	•

3.3. Spatial and Seasonal Variation in Species Richness, Abundance and Diversity

Mean species richness, abundance, and diversity varied by study site and season (Figure 4). Mean species richness at the estuary was higher during the warm-rainy than the cold-dry season, whereas no differences were found between the two seasons at the freshwater site ($F = 3.724$, $P < 0.05$). The mean abundances of fishes were higher during the warm-rainy season at both the freshwater and estuary, with the lowest value during the cold-dry season at the estuary ($F = 2.182$, $P < 0.05$). Diversities also varied between study sites. A higher diversity was found at the estuary than at the freshwater site during the warm-rainy season only ($F = 0.054$, $P < 0.05$).

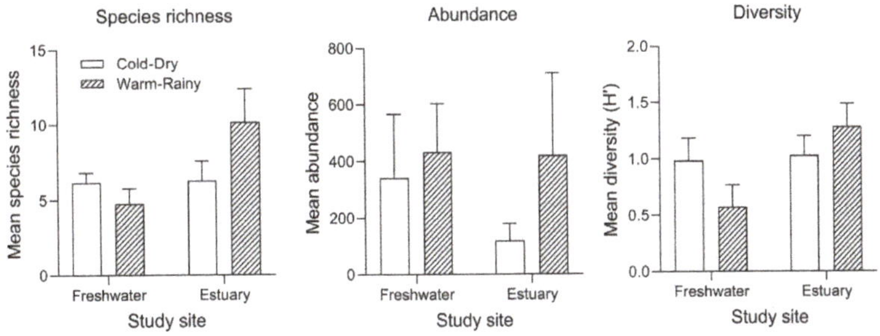

Figure 4. Variations in mean species richness, abundance, and diversity of fish assemblages with study site and season in the lower Nakdong River, South Korea.

The two-way ANOVA confirmed diversity patterns, in that there were significant effects of the site, but not season (Table 2). The two-way interactions were also significant for species richness and diversity, with the exception of abundance (Table 2). Split file ANOVAs showed significant trends of species richness and diversity with site for the warm-rainy season ($P < 0.05$). No significance for any community variable was observed between sites during the cold-dry season ($P > 0.05$). Abundance differences were significant between the freshwater and estuarine site during the cold-dry season only. Within-site seasonal comparisons showed a significant difference for abundance at the estuary habitat and diversity at the freshwater site ($P < 0.05$).

Table 2. Results of two-way analysis of variance (ANOVA) on the number of species, abundance, and diversity of fish assemblages in the lower Nakdong River, South Korea; bold face indicates statistical significance at $P < 0.05$.

Source	df	Species Richness		Abundance		Diversity	
		F	*P*	*F*	*P*	*F*	*P*
Site	1	9.056	**0.004**	7.775	**0.007**	7.840	**0.007**
Season	1	1.815	0.184	4.000	0.051	0.348	0.558
Site × Season	1	8.262	**0.006**	0.823	0.369	6.093	**0.017**
Residual	53						

Mean abundances of the eight dominant species varied seasonally within each habitat (Figure 5). In the freshwater habitat, the mean abundances of *He. labeo* and *Ps. escocinus* were higher in the warm-rainy season, while *Ch. erythropterus* was abundant during cold-dry season (ANOVA, $P < 0.05$). The estuarine fishes showed tendencies of higher abundance during the warm-rainy season only ($P < 0.05$). No seasonal differences of mean abundances were observed for *Ta. nitidus* in freshwater habitats, and for *Nu. nuchalis* in the estuary region (ANOVA, $P > 0.05$).

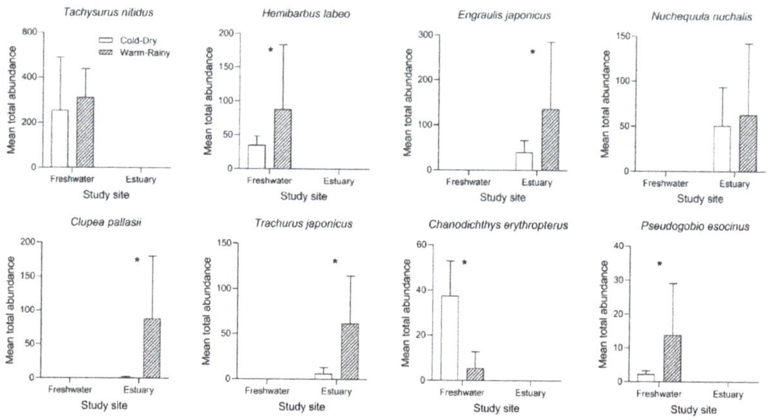

Figure 5. Seasonal variations in mean abundance of 8 common fish species within each freshwater and estuary habitats in the lower Nakdong River, South Korea; asterisks indicate statistical significance at $P < 0.05$ between seasons within each site.

3.4. Fish Assemblage Structure

PERMANOVA tests revealed fish assemblages were significantly associated with study site and season, but not station (nested within study site), with the COV of site being the highest, indicating the strongest factor determining variation within samples (Table 3). A statistically significant two-way interaction between site and season was also observed (Table 3). Pairwise comparisons of site and season showed evidence of significant differences in fish assemblage structure between freshwater and estuarine site within each season, and also between cold-dry and warm-rainy season with each of study site (PERMANOVA pairwise tests, all $P < 0.05$).

Table 3. Mean squares (MS), pseudo-F ratios, significance levels (P), and components of variation (COV) for permutation multivariate analyses of variance (PERMANOVA) tests using Bray–Curtis similarity matrices from abundance of fish assemblages showing differences in site (Si), station (St, nested within Si), season (Se), and interactions terms in the lower Nakdong River, South Korea; bold letters indicate significance at $P < 0.05$.

Source	df	MS	Pseudo-F	*P*	COV
Si	1	70860.0	30.243	**0.007**	50.884
Se	1	7596.3	6.333	**0.025**	15.548
St (Si)	3	2353.8	1.227	0.131	6.207
Si × Se	1	7611.4	6.346	**0.023**	22.014
St (Si) × Se	3	1181.2	0.616	0.988	−11.419
Residuals	47	1918.3			43.798

The non-metric MDS ordination of similarity of mean fish assemblages depicted a clear visual difference between freshwater and estuarine site along with the nMDS ordination horizontally, with the former and the latter site lying right and left side of the plot, respectively (Figure 6). Conversely, the points during both cold-dry and warm-rainy seasons at the estuarine site were interspersed throughout the nMDS plot. At the freshwater site, the points for the cold-dry season lie toward the bottom of the plot, while those for the warm-rainy season were at the upper plot area.

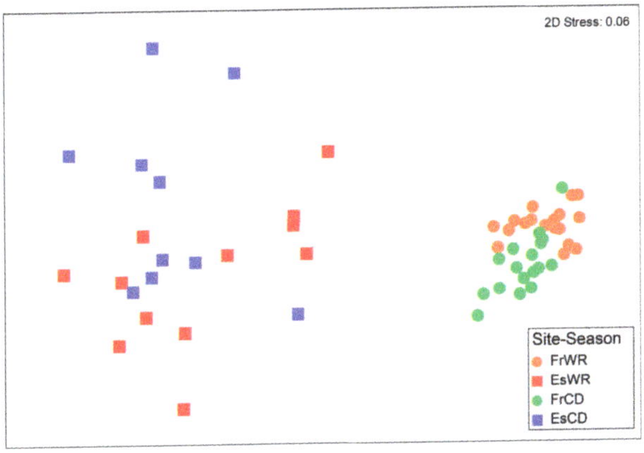

Figure 6. Non-metric multidimensional scaling (nMDS) for fish assemblages from the two study sites during cold-dry and warm-rainy seasons in the lower Nakdong River, South Korea.; Fr = Freshwater, Es = Estuary; WR = warm-rainy season, CD = cold-dry season.

Canonical analyses on principal coordinates were performed on significant interactions as a further test on PERMANOVA analyses. The CAP plot for the site–season interaction showed strong evidence for the factor-indicating group separation between sites, and between seasons within each site (Figure 7). Seven fish species were key in separating the estuarine site from the freshwater site, and four species characterized the fish assemblages in freshwater habitats. *Co. nasus* had an intermediate contribution on both sites during the warm-rainy season (Figure 7). Clear seasonal differences in fish assemblages were found at the estuarine site, showing a strong contribution for six species on the warm-rainy season and only a contribution from *Nu. nuchalis* to trends in the cold-dry season. Weak seasonal classifications in fish assemblages were evident at the freshwater site, with weak trends of species contributions on each season (Figure 7).

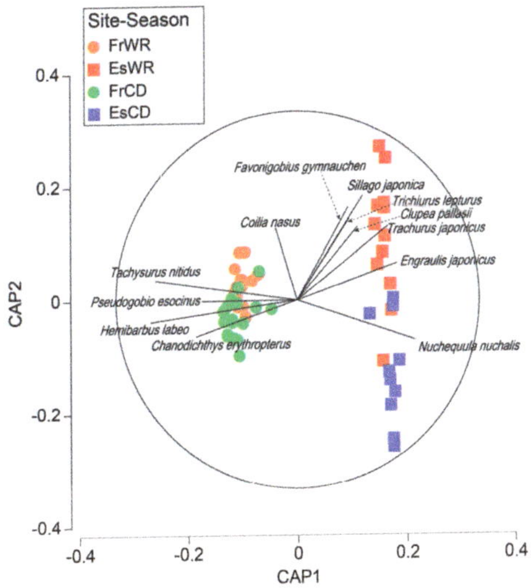

Figure 7. Ordination plots for canonical analysis of principal coordinates of fish assemblage site–season interactions in the lower Nakdong River, South Korea; Fr = Freshwater, Es = Estuary; WR = warm-rainy season, CD = cold-dry season.

3.5. Ecological Guild and Habitat Type of Fish Assemblages

The analysis of ecological guilds by the number of species within each site across each season showed a high influence of 12 and 13 freshwater species (FW) during the cold-dry and the warm-rainy season, respectively. Collectively, two diadromous migrants (DM) and one estuarine resident (ER) were recorded at the freshwater site during both seasons (Figure 8). In the estuarine site, 13 and 20 marine occasional residents (MV) occurred during the cold-dry and warm-rainy seasons, respectively. No freshwater species were found at this region. All ecological guilds occurred in fishway samples, with dominant freshwater and marine occasional residents (Table 1), indicating that fishways play a role in connecting freshwater and estuary habitats. Marine juveniles (MJ) were only recorded at fishways and the estuary site, with the greatest number recorded in the latter. Freshwater habitats comprised mostly of pelagic species, with the benthic-species number higher at the estuary and fishway regions.

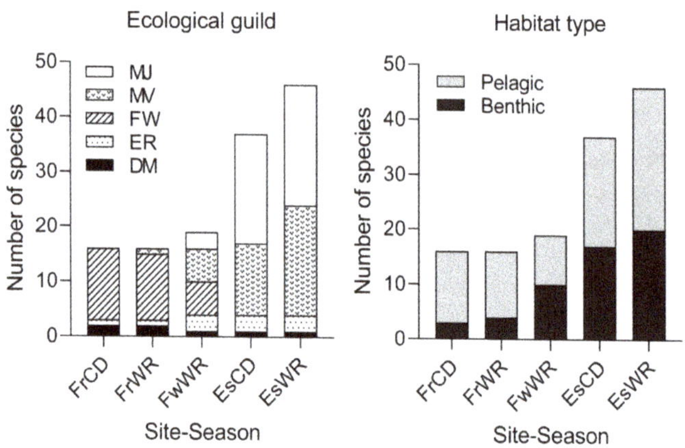

Figure 8. The number of species by ecological guild and habitat type in freshwater (Fr), fishway (Fw) and estuary (Es) sites during cold-dry (CD) and warm-rainy (WR) seasons; ecological guilds were estuarine residents (ER), freshwater species (FW), marine occasional visitors (MV), marine juveniles (MJ), and diadromous migrants (DM).

4. Discussion

A total of 18 and 63 species were collected from freshwater and estuary habitats, respectively, whereas 19 fish species were recorded from fishways. The number of species and diversity was considerably higher at estuary stations. Higher species richness in estuarine regions, compared with freshwater or blackish habitats, has been reported worldwide [41,42]. De Moura et al. [42] indicated that species richness was significantly higher in channels linking estuaries to freshwater ecosystems, due to the regular inflow of saltwater, allowing access for several marine species. Additionally, higher species richness, especially toward estuary mouths, is strongly influenced by marine processes, supporting a greater number of species [43]. Processes such as salinity gradients within the estuary have been shown to strongly influence fish species richness [44]. Such influences, however, may not be general in fish assemblages for all estuaries. Species richness has also been shown to be influenced by regional and local processes affecting colonization, such as processes affecting connectivity between estuaries and the adjacent marine habitats [45].

Of the 63 fish species recorded in the estuarine stations, five were numerically dominant (*En. japonicus*, *Nu. nuchalis*, *Cl. pallasii*, *Tr. japonicus* and *Fa. Gymnauchen*). Estuarine fishes were predominantly juveniles of marine fishes, indicating the importance of the estuary as a nursery [27,46]. The dominance of juvenile fishes in our observations is in general agreement with other studies worldwide (e.g., [47–50]), as well as in Korea (e.g., [26,27,51]). The greater abundance of juveniles observed in this study indicated the strong dependence of species on the estuary for shelter, survival, and refuge from predators during early stages of their life cycle [48]. As an additional importance of estuaries, most of the marine species that use the estuary as a nursery ground have commercial and recreational value. Among the dominant marine juveniles, *Cl. pallasii*, *Pa. olivaceus*, *Tr. japonicus* and *Tr. lepturus* are the most important fishery resources in Korea [52].

Eighteen fish species were collected in the upper part of estuary weir, of which 14 species were freshwater residents and two species diadromous migrants (*Co. nasus* and *Mu. cephalus*). Among the dominant freshwater species, *Ch. erythropterus* and *He. labeo* are endemic to the Nakdong River [53,54], and *Ta. nitidus* comprises a minor group of freshwater fishes in the Nakdong River [55]. It is worth noting that *Opsariichthys uncirostris*, the common species in freshwater and estuary habitats of the lower Nakdong River [10], was not a major species observed in this study. The low occurrence of *Op. uncirostris* is likely due to extensive dredging between 2009 and 2011 in the lower Nakdong River. The river dredging caused deteriorated water quality and compromised critical habitats, causing the observed low numbers of aquatic organisms [56]. Highly tolerant species to such disturbances, including *Ta. Nitidus*, were abundant in the current study, unlike more sensitive species, such as *Op. uncirostris* [57].

The estuarine weir of the Nakdong River was constructed in 1987, despite many arguments against it, due to the many potential negative impacts on the ecosystem [58,59]. One such impact may be on the changes of fish assemblage structures. The major change in fish assemblage is a decrease of total species number, especially freshwater fishes after the construction of estuary weir [60]. Changes in fish functional categories were also observed from demersal to pelagic in estuary fish assemblages [20]. In addition, migratory fishes also have shown dramatical reduction of their upstream migration route and relative abundance due to blockage of their passage by the estuary weir [60]. Thus, the role of fishways in determining linkages between freshwater and estuarine habitats may be particularly significant for shaping fish communities, especially communities of diadromous migrants.

The results of this study show similar proportions of fish species within freshwater, marine, and stenohaline (estuarine residents and diadromous migrants) guilds in the fishway samples. Fishways were constructed for the provision of passages for animals migrating from estuary to freshwater habitats. Relatively few diadromous species, however, were found in this study, despite the fishways, even with many diadromous species present in the lower reaches of the Nakdong River [10,61]. Because of the obstruction of the estuary weir in fishes' migration into the upper river, several studies have reported reduced migratory fishes in the Nakdong River estuary [10]. Yang et al. [61] reported fishways in the weir of Nakdong River estuary to have limited use, based on the relatively low species richness of anadromous fishes upriver from the structure. In addition, reduced freshwater flow caused by low precipitation can also exacerbate negative impacts on migratory fishes, due to fishways from weir construction becoming of even less efficiency in the provision of adequate passage [62,63]. Effective alternatives for providing fish passage, such as new fish passage designs, implementing recovery of affected habitats, and implementing adaptive management practices, are urgent and necessary for areas in Korea where fishways exist or are planned.

Overall diversity patterns of fish assemblages did not show significant seasonal patterns. A higher fish abundance, however, in the estuarine site was evident during the warm-rainy season. Seasonal variations in fish abundance are mainly due to the presence of species utilizing the Nakdong River estuary during the warm-rainy season for reproduction. For example, the increase in the abundance of *Cl. pallasii*, *En. Japonicus*, and *Tr. japonicus* occurred mainly during the warm-rainy season for *Cl. pallasii* following their spawning periods [64], and during the spring-early summer for *En. japonicus* and *Tr. japonicus* [65,66]. Those same species are also abundant at adjacent near-coastal waters of southeastern Korea during the spring and summer [56,67]. One may thus conclude that the abundances of those species during the warm-rainy season are likely due to the nursery grounds provided by the estuary, whereas the consistent occurrence of *Nu. nucalis* during all seasons suggests that estuaries are the main habitat for their entire life cycle [20].

5. Conclusions

In conclusion, the lower reaches of the Nakdong River are an important ecosystem for a diverse array of fish species. Of the 78 species sampled in this study, 57 and 14 originated from marine and freshwater areas, respectively, even with limited access to their habitats by the estuary weir. Most of the species sampled in estuaries were exclusively represented by juveniles, highlighting the importance of this ecosystem as a nursery ground. As in many estuaries across the world, many of the species observed in this study were occasional residents. The effect of obstructions, such as the estuary weir, on their success was reflected by the reduction of species richness in areas upriver of the weir than in the previous observation [60]. Studies such as this, investigating the dynamics of species assemblage in lowland river systems impacted by physical obstructions, are critical for the conservation efforts of habitats supporting ecologically and economically important species.

Supplementary Materials: The following are available online at http://www.mdpi.com/2077-1312/8/7/496/s1, Table S1: List of species including Families, Scientific names, Common names and their abundances (N) in freshwater and estuary sites, and total abundance as well as percentages (%N).

Author Contributions: All authors have read and agree to the published version of the manuscript. J.M.P.; methodology, validation, formal analysis and writing—original draft preparation, R.R.; writing—review & editing; H.H.J.; conceptualization, H.C.C.; investigation and data curation.

Funding: This research was funded by the Korea Institute of Ocean Science & Technology [grant numbers PE99813].

Acknowledgments: We are grateful to Ki Mun Nam and Dong Jin Lee for assistance with samplings and data analyses. Field surveys were conducted in accordance with the approval of "Research & Training Fishery" in the Ministry of Ocean and Fisheries, Korea.

Conflicts of Interest: The authors declare no conflict of interest.

References

1. Day, J.W., Jr.; Kemp, W.M.; Yáñez-Arancibia, A.; Crump, B.C. *Estuarine Ecology*; John Wiley & Sons: Hoboken, NJ, USA, 2012.
2. Dolbeth, M.; Cardoso, P.G.; Ferreira, S.M.; Verdelhos, T.; Raffaelli, D.; Pardal, M.A. Anthropogenic and natural disturbance effects on a macrobenthic estuarine community over a 10-year period. *Mar. Pollut. Bull.* **2007**, *54*, 576–585. [CrossRef] [PubMed]
3. Gaonkar, U.V.; Sivadas, S.K.; Ingole, B.S. Effect of tropical rainfall in structuring the macrobenthic community of Mandovi estuary, west coast of India. *J. Mar. Biol. Assoc. U. K.* **2013**, *93*, 1727–1738. [CrossRef]
4. Schuchardt, B.; Haesloop, U.; Schirmer, M. The tidal freshwater reach of the Weser estuary: Riverine or estuarine? *Neth. J. Aquat. Ecol.* **1993**, *27*, 215–226. [CrossRef]
5. Cardoso, P.G.; Raffaelli, D.; Lillebo, A.I.; Verdelhos, T.; Pardal, M.A. The impact of extreme flooding events and anthropogenic stressors on the macrobenthic communities' dynamics. *Estuar. Coast. Shelf Sci.* **2008**, *76*, 553–565. [CrossRef]
6. McDowall, R.M. The evolution of diadromy in fishes (revisited) and its place in phylogenetic analysis. *Rev. Fish. Biol. Fish.* **1997**, *7*, 443–462. [CrossRef]
7. Kim, I.S.; Choi, Y.; Lee, C.; Lee, Y.; Kim, B.; Kim, J.; Choi, Y.; Lee, Y.; Kim, B. *Illustrated Book of Korean Fishes*; Kyo-Hak: Seoul, Korea, 2005.
8. Froese, R.; Pauly, D. FishBase. World Wide Web Electronic Publication. Version (12/2019). Available online: https://www.fishbase.org/ (accessed on 16 April 2020).
9. Hwang, J.-S. The fish migration through fishlock at Youngam lake. *J. Korea Water Resour. Assoc.* **2000**, *33*, 561–568.
10. Kang, E.-J.; Yang, H.; Lee, H.-H.; Kim, K.-S.; Kim, C.-H. Characteristics of fish fauna collected from near estuaries bank and fish-way on the bank of Naktong river. *Korean J. Ichthyol.* **2012**, *24*, 201–219.
11. Jackson, D.A.; Peres-Neto, P.R.; Olden, J.D. What controls who is where in freshwater fish communities-the roles of biotic, abiotic, and spatial factors. *Can. J. Fish. Aquat. Sci.* **2001**, *58*, 157–170. [CrossRef]
12. Joy, M.K.; Death, R.G. Control of freshwater fish and crayfish community structure in Taranaki, New Zealand: Dams, diadromy or habitat structure? *Freshw. Biol.* **2001**, *46*, 417–429. [CrossRef]

13. Gillanders, B.M.; Kingsford, M.J. Impact of changes in flow of freshwater on estuarine and open coastal habitats and the associated organisms. *Oceanogr. Mar. Biol. An Annu. Rev.* **2002**, *40*, 233–309.
14. Gehrke, P.C.; Gilligan, D.M.; Barwick, M. Changes in fish communities of the Shoalhaven River 20 years after construction of Tallowa Dam, Australia. *River Res. Appl.* **2002**, *18*, 265–286. [CrossRef]
15. Rolls, R.J. The role of life-history and location of barriers to migration in the spatial distribution and conservation of fish assemblages in a coastal river system. *Biol. Conserv.* **2011**, *144*, 339–349. [CrossRef]
16. Han, M.; Fukushima, M.; Kameyama, S.; Fukushima, T.; Matsushita, B. How do dams affect freshwater fish distributions in Japan? Statistical analysis of native and nonnative species with various life histories. *Ecol. Res.* **2008**, *23*, 735–743. [CrossRef]
17. Jeon, S. Studies on the fish fauna of the estuary area of Naktong river, Korea. *Bull. Kor. Assoc. Conser. Nat.* **1987**, *9*, 77–90.
18. Flemer, D.A.; Champ, M.A. What is the future fate of estuaries given nutrient over-enrichment, freshwater diversion and low flows? *Mar. Pollut. Bull.* **2006**, *52*, 247–258. [CrossRef] [PubMed]
19. Jang, I.-K.; Kim, C.-H. A study on the changes of the molluscan and crustacean fauna after the construction of the Naktong estuary barrage. *Korean J. Fish. Aquat. Sci.* **1992**, *25*, 265–281.
20. Kwak, S.N.; Huh, S.-H. Changes in species composition of fishes in the Nakdong River Estuary. *Korean J. Fish. Aquat. Sci.* **2003**, *36*, 129–135.
21. Park, S.; Lee, J.; Choi, J.U.; Heo, N.; An, S. Study on the long-term changes in water quality and benthic ecology and evaluation on effect of the barrage in Nakdong River estuary. *J. Wetl. Res.* **2016**, *18*, 58–67. [CrossRef]
22. Verhelst, P.; Buysse, D.; Reubens, J.; Pauwels, I.; Aelterman, B.; Van Hoey, S.; Goethals, P.; Coeck, J.; Moens, T.; Mouton, A. Downstream migration of European eel (Anguilla anguilla L.) in an anthropogenically regulated freshwater system: Implications for management. *Fish. Res.* **2018**, *199*, 252–262. [CrossRef]
23. Potter, I.C.; Claridge, P.N.; Warwick, R.M. Consistency of Seasonal-Changes in an Estuarine Fish Assemblage. *Mar. Ecol. Prog. Ser.* **1986**, *32*, 217–228. [CrossRef]
24. Hood, W.G. Indirect environmental effects of dikes on estuarine tidal channels: Thinking outside of the dike for habitat restoration and monitoring. *Estuaries* **2004**, *27*, 273–282. [CrossRef]
25. Zampatti, B.P.; Bice, C.M.; Jennings, P.R. Temporal variability in fish assemblage structure and recruitment in a freshwater-deprived estuary: The Coorong, Australia. *Mar. Freshw. Res.* **2010**, *61*, 1298–1312. [CrossRef]
26. Hwang, S.W.; Lee, T.W.; Hwang, H.B.; Do Hwang, S. Temporal occurrence patterns and habitat use of fish in the Mangyeong Estuary on the western coast of Korea in 2003, before dike enclosure. *Ocean Sci. J.* **2011**, *46*, 307–314. [CrossRef]
27. Park, J.M.; Huh, S.H.; Baeck, G.W. Temporal variations of fish assemblage in the surf zone of the Nakdong River Estuary, southeastern Korea. *Anim. Cells Syst.* **2015**, *19*, 350–358. [CrossRef]
28. Mariani, S. Can spatial distribution of ichthyofauna describe marine influence on coastal lagoons? A central Mediterranean case study. *Estuar. Coast. Shelf Sci.* **2001**, *52*, 261–267. [CrossRef]
29. Ayvazian, S.; Hyndes, G. Surf-zone fish assemblages in south-western Australia: Do adjacent nearshore habitats and the warm Leeuwin Current influence the characteristics of the fish fauna? *Mar. Biol.* **1995**, *122*, 527–536. [CrossRef]
30. Akin, S.; Buhan, E.; Winemiller, K.O.; Yimaz, H. Fish assemblage structure of Koycegiz Lagoon-Estuary, Turkey: Spatial and temporal distribution patterns in relation to environmental variation. *Estuar. Coast. Shelf Sci.* **2005**, *64*, 671–684. [CrossRef]
31. Potter, I.C.; Bird, D.J.; Claridge, P.; Clarke, K.; Hyndes, G.; Newton, L. Fish fauna of the Severn Estuary. Are there long-term changes in abundance and species composition and are the recruitment patterns of the main marine species correlated? *J. Exp. Mar. Biol. Ecol.* **2001**, *258*, 15–37. [CrossRef]
32. Contente, R.F.; Stefanoni, M.F.; Spach, H.L. Fish assemblage structure in an estuary of the Atlantic Forest biodiversity hotspot (southern Brazil). *Ichthyol. Res.* **2011**, *58*, 38–50. [CrossRef]
33. Fausch, K.D.; Torgersen, C.E.; Baxter, C.V.; Li, H.W. Landscapes to riverscapes: Bridging the gap between research and conservation of stream fishes: A continuous view of the river is needed to understand how processes interacting among scales set the context for stream fishes and their habitat. *BioScience* **2002**, *52*, 483–498. [CrossRef]
34. Durance, I.; Lepichon, C.; Ormerod, S.J. Recognizing the importance of scale in the ecology and management of riverine fish. *River Res. Appl.* **2006**, *22*, 1143–1152. [CrossRef]

35. Chyung, M.K. *The Fishes of Korea*; Il Ji Sa Publishing Co.: Seoul, Korea, 1977.
36. Shannon, C.; Weaver, W. *The Mathematical Theory of Communication*; Illinois University Press: Urbana, IL, USA, 1949.
37. Sokal, R.R.; Rohlf, F.J. *Biometry: The Principles and Practice of Statistics in Biological Research*; Freeman New York: New York, NY, USA, 1981.
38. Clarke, K.R.; Somerfield, P.J.; Chapman, M.G. On resemblance measures for ecological studies, including taxonomic dissimilarities and a zero-adjusted Bray-Curtis coefficient for denuded assemblages. *J. Exp. Mar. Biol. Ecol.* **2006**, *330*, 55–80. [CrossRef]
39. Anderson, M.; Gorley, R.; Clarke, K. *PERMANOVA+ for PRIMER: Guide to Software and Statistical Methods*; PRIMER-E Plymouth Marine Laboratory: Plymouth, UK, 2008.
40. Clarke, K.; Gorley, R. *PRIMER V7: User Manual/Tutorial*; PRIMER-E: Plymouth, UK, 2015.
41. Martino, E.J.; Able, K.W. Fish assemblages across the marine to low salinity transition zone of a temperate estuary. *Estuar. Coast. Shelf Sci.* **2003**, *56*, 969–987. [CrossRef]
42. De Moura, P.M.; Vieira, J.P.; Garcia, A.M. Fish abundance and species richness across an estuarine-freshwater ecosystem in the Neotropics. *Hydrobiologia* **2012**, *696*, 107–122. [CrossRef]
43. Yoklavich, M.M.; Cailliet, G.M.; Barry, J.P.; Ambrose, D.A.; Antrim, B.S. Temporal and Spatial Patterns in Abundance and Diversity of Fish Assemblages in Elkhorn Slough, California. *Estuaries* **1991**, *14*, 465–480. [CrossRef]
44. Sosa-Lopez, A.; Mouillot, D.; Ramos-Miranda, J.; Flores-Hernandez, D.; Chi, T.D. Fish species richness decreases with salinity in tropical coastal lagoons. *J. Biogeogr.* **2007**, *34*, 52–61. [CrossRef]
45. Vasconcelos, R.P.; Henriques, S.; Franca, S.; Pasquaud, S.; Cardoso, I.; Laborde, M.; Cabral, H.N. Global patterns and predictors of fish species richness in estuaries. *J. Anim. Ecol.* **2015**, *84*, 1331–1341. [CrossRef]
46. Martinho, F.; Leitão, R.; Neto, J.M.; Cabral, H.N.; Marques, J.C.; Pardal, M.A. The use of nursery areas by juvenile fish in a temperate estuary, Portugal. *Hydrobiologia* **2007**, *587*, 281–290. [CrossRef]
47. Ikejima, K.; Tongnunui, P.; Medej, T.; Taniuchi, T. Juvenile and small fishes in a mangrove estuary in Trang province, Thailand: Seasonal and habitat differences. *Estuar. Coast. Shelf Sci.* **2003**, *56*, 447–457. [CrossRef]
48. Vidy, G.; Darboe, F.S.; Mbye, E.M. Juvenile fish assemblages in the creeks of the Gambia Estuary. *Aquat. Living Resour.* **2004**, *17*, 56–64. [CrossRef]
49. Plavan, A.A.; Passadore, C.; Gimenez, L. Fish assemblage in a temperate estuary on the uruguayan coast: Seasonal variation and environmental influence. *Braz. J. Oceanogr.* **2010**, *58*, 299–314. [CrossRef]
50. Weitkamp, L.A.; Bentley, P.J.; Litz, M.N.C. Seasonal and interannual variation in juvenile salmonids and associated fish assemblage in open waters of the lower Columbia River estuary. *Fish. Bull.* **2012**, *110*, 426–450.
51. Hwang, S.-W.; Hwang, H.-B.; Noh, H.-S.; Lee, T.-W. Seasonal variation in species composition of fish collected by a bag net in the Geum River estuary, Korea. *Korean J. Fish. Aquat. Sci.* **2005**, *38*, 39–54. [CrossRef]
52. Kim, Y.; Han, K.; Kang, C.; Kim, J. *Commercial Fishes of the Coastal and Offshore Waters in Korea*; Hangel: Busan. Korea, 2004.
53. Sung, N.J.; Lee, E.H.; Ha, B.S. Studies on the food from fresh water fish-3. Composition in amino acids and nucleotides of the dorsal muscle of some fresh water fishes, Sweet Fish, Cornet Fish, Mandarin Fish and Read Fish. *J. Korean Soc. Food Sci. Nutr.* **1984**, *13*, 163–168.
54. Yoon, J.D.; Kim, J.H.; In, D.S.; Yu, J.J.; Hur, M.; Chang, K.H.; Jang, M.H. Evaluation of movement pattern of *Erythroculter erythropterus* inhabit in the mid-lower part of Nakdong River using acoustic telemetry. *Korean J. Ecol. Environ.* **2012**, *45*, 403–411. [CrossRef]
55. Jang, S.-H.; Zhang, C.-I.; Na, J.-H.; Lee, J.-H. Analysis of trophic structures and energy flows in aquatic ecosystem of the lower reaches of the Nakdong River. *Korean J. Environ. Biol.* **2008**, *26*, 292–302.
56. Jeong, J.M.; Park, J.M.; Huh, S.H.; Ye, S.J.; Kim, H.J.; Baeck, G.W. Seasonal variation in the species composition of fish assemblages in the coastal waters off Gadeok-do, South sea, Korea. *Korean J. Fish. Aquat. Sci.* **2013**, *46*, 948–956. [CrossRef]
57. Sato, M.; Kawaguchi, Y.; Yamanaka, H.; Okunaka, T.; Nakajima, J.; Mitani, Y.; Shimatani, Y.; Mukai, T.; Onikura, N. Predicting the spatial distribution of the invasive piscivorous chub (*Opsariichthys uncirostris uncirostris*) in the irrigation ditches of Kyushu, Japan: A tool for the risk management of biological invasions. *Biol. Invasions* **2010**, *12*, 3677–3686. [CrossRef]
58. Williams, J.R.; Dellapenna, T.M.; Lee, G.H. Shifts in depositional environments as a natural response to anthropogenic alterations: Nakdong Estuary, South Korea. *Mar. Geol.* **2013**, *343*, 47–61. [CrossRef]

59. Park, S.R.; Kim, J.H.; Kang, C.K.; An, S.; Chung, I.K.; Kim, J.H.; Lee, K.S. Current status and ecological roles of *Zostera marina* after recovery from large-scale reclamation in the Nakdong River estuary, Korea. *Estuar. Coast. Shelf Sci.* **2009**, *81*, 38–48. [CrossRef]
60. Ju, K.J. Changes in fish fauna and evaluation of migratory fishes before and after the construction of the Nakdong River Estuary weir. *Nakdong River Res.* **2005**, 2–14.
61. Yang, H.J.; Kim, K.H.; Kum, J.D. The fish fauna and migration of the fishes in the fish way of the Nakdong River Mouth Dam. *J. Ecol. Environ.* **2001**, *34*, 251–258.
62. O'Brien, T. Overcoming Physical Barriers to Fish Migration. In Proceedings of the Fish Movement and Migration, Australian Society for Fish Biology Workshop Proceedings, Bendigo, Victoria, Australia, 27–29 September 1999; pp. 129–134.
63. Gillson, J.; Scandol, J.; Suthers, I. Estuarine gillnet fishery catch rates decline during drought in eastern Australia. *Fish. Res.* **2009**, *99*, 26–37. [CrossRef]
64. Hoshikawa, H.; Kuwahara, H.; Tajima, K.; Kawai, T.; Kaneta, T.; Tsuda, F. Characteristics of a pacific herring *Clupea pallasii* spawning bed off Minedomari, Hokkaido, Japan. *Fish. Sci.* **2004**, *70*, 772–779. [CrossRef]
65. Funamoto, T.; Aoki, I.; Wada, Y. Reproductive characteristics of Japanese anchovy, *Engraulis japonicus*, in two bays of Japan. *Fish. Res.* **2004**, *70*, 71–81. [CrossRef]
66. Cha, H.K.; Lee, J.B.; Kang, S.K.; Chang, D.S.; Choi, J.H. Reproduction of the jack mackerel, Trachurus japonicus Temminck et Schlegel in the coastal waters around Jeju Island, Korea: Maturation and spawning. *J. Korean Soc. Fish. Technol.* **2009**, *45*, 243–250. [CrossRef]
67. Park, J.M.; Baeck, G.W.; Huh, S.H. Variations in demersal fish assemblage in the southern coast of East Sea, Korea. *Korean J. Fish. Aquat. Sci.* **2018**, *51*, 426–434. [CrossRef]

Journal of
Marine Science and Engineering

MDPI

Article

Population Genetic Diversity of Two Marine Gobies (Gobiiformes: Gobiidae) from the North-Eastern Atlantic and the Mediterranean Sea

Katarína Čekovská [1], Radek Šanda [2], Kristýna Eliášová [3,4], Marcelo Kovačić [5], Stamatis Zogaris [6], Anna Maria Pappalardo [7], Tereza Soukupová [1] and Jasna Vukić [1,*]

1 Department of Ecology, Faculty of Science, Charles University, Viničná 7, 128 44 Prague, Czech Republic; chalupek@natur.cuni.cz (K.Č.); terka.slamova@gmail.com (T.S.)
2 Department of Zoology, National Museum, Václavské nám. 68, 110 00 Prague, Czech Republic; radek.sanda@nm.cz
3 Department of Zoology, Faculty of Science, Charles University, Viničná 7, 128 44 Prague, Czech Republic; kristyna.eliasova@natur.cuni.cz
4 Department of Animal Science and Food Processing, Faculty of Tropical AgriSciences, Czech University of Life Sciences Prague, Kamýcka 961/129, 165 00 Prague, Czech Republic
5 Natural History Museum Rijeka, Lorenzov prolaz 1, HR-51000 Rijeka, Croatia; marcelo@prirodoslovni.com
6 Institute of Marine Biological Sciences and Inland Waters, Hellenic Centre for Marine Research, Athinon-Souniou Ave., 19013 Anavissos, Greece; zogaris@hcmr.gr
7 Department of Biological, Geological and Environmental Sciences, University of Catania, Via Androne 81, 95124 Catania, Italy; pappalam@unict.it
* Correspondence: jasna.vukicova@natur.cuni.cz

Received: 2 September 2020; Accepted: 10 October 2020; Published: 13 October 2020

check for updates

Abstract: Gobies (Gobiiformes: Gobiidae) are the most species-rich family of fishes in general, and the most abundant fish group in the European seas. Nonetheless, our knowledge on many aspects of their biology, including the population genetic diversity, is poor. Although barriers to gene flow are less apparent in the marine environment, the ocean is not a continuous habitat, as has been shown by studies on population genetics of various marine biota. For the first time, European marine goby species which cannot be collected by common fishery techniques were studied. The population genetic structure of two epibenthic species, *Gobius geniporus* and *Gobius cruentatus*, from seven localities across their distribution ranges was assessed, using one mitochondrial (cytochrome b) and one nuclear gene (first intron of ribosomal protein gene S7). Our results showed that there is a great diversity of haplotypes of mitochondrial gene cytochrome b in both species at all localities. Global fixation indices (F_{ST}) indicated a great differentiation of populations in both studied gobies. Our results did not show a geographic subdivision to individual populations. Instead, the data correspond with the model of migration which allow divergence and recurrent migration from the ancestral population. The estimated migration routes coincide with the main currents in the studied area. This matches well the biology of the studied species, with adults exhibiting only short-distance movements and planktonic larval stages.

Keywords: benthic fish; molecular tools; cytochrome b; ribosomal protein gene S7; *Gobius cruentatus*; *Gobius geniporus*; Mediterranean Sea; genetic structure

1. Introduction

Despite the fact that the seas and oceans are interconnected, marine organisms can show a strong genetic differentiation [1]. Several phylogeographical studies have shown that even the ocean is a fragmented environment. A spatial genetic structure has been discovered in different marine organisms, e.g., in sea-grasses [2], sponges [3], mollusks [4], sea cucumbers [5], sea urchins [6], crustaceans [7–9] or fishes [10–12]. However, in contrast to terrestrial and freshwater ecosystems, the delimitation of individual populations in the ocean is not an easy task because barriers to gene flow are far less apparent in the marine environment [13]. Some of the driving forces of the genetic variability of marine species are oceanographic barriers (e.g., direction of currents, presence of straits, extent of different type of habitats, temperature and salinity zonation), limited dispersal capabilities of species, isolation by distance and geological history of the area. On the other hand, factors such as hydrodynamics, long duration of the larval pelagic stage, or migratory behaviour of adults are most commonly responsible for the genetic homogenization of the population [1,14]. Final scenario of genetic partitioning can thus be a consequence of interaction of more factors. Understanding these processes is crucial for marine phylogeographical investigations, for species conservation and management of marine resources [15].

The Mediterranean Sea is a small enclosed basin, connected with the Atlantic Ocean by the Strait of Gibraltar and with the Black Sea through the Bosporus. It is subdivided into several deep subbasins separated by shallow sills. The major water exchange occurs with the Atlantic Ocean, and is strongly affected by climate [16]. Cool surface water inflows from the Atlantic Ocean to the Mediterranean Sea, while warm and more saline subsurface Mediterranean water outflows to the Atlantic Ocean [16].

Although they are adjacent and were connected during most of their past, the Mediterranean Sea and the Atlantic Ocean had partly dissimilar geological histories. Driven by a combination of climatic and tectonic forces, the Messinian Salinity Crisis (MSC, 5.97 to 5.33 million years ago, Mya) was one of the major events, which impacted geological history, and consequently the biota of the Mediterranean Sea [17,18]. It is widely accepted that at the onset of the MSC, the Mediterranean Sea became disconnected from the world ocean, and, as a consequence of evaporation, suffered a great water level drawdown [18,19]. This led to a severe change of environmental conditions in the Mediterranean Sea. However, the scenarios of the fate of the Mediterranean Sea at that time greatly differ (from almost complete desiccation of the sea to the existence of the deepwater marine environment in the first phase of the MSC [17,18,20]). The second phase of the MSC was characterized by fluctuations of environmental conditions due to the repeated connection with the Paratethys [17,21]. Accordingly with the various scenarios about the form of the Mediterranean Sea during the MSC, various scenarios about the fate of the ichthyofauna in the Mediterranean Sea were proposed, ranging from the extinction [22] to survival, possibly in refugia; the latter was corroborated by the findings of fossils of marine fishes [21,23]. The MSC ended by the opening of Gibraltar and refilling of the Mediterranean Sea by Atlantic waters in an event known as the Zanclean flooding [24] and by the onset of stable marine conditions [17].

Another event with a strong impact on the diversity and distribution of the extant species in the Mediterranean Sea and the Atlantic was the Pleistocene glaciation (2.6 Mya to 11,600 before present, BP) [25]. During the most recent ice ages the growth and decay of ice masses drove the world sea-level fluctuations in the order of 10's to over 100 m on the time scales of 100's to 10,000 years, and ranging from the sea level several meters higher than present to more than 100 m below the present level [26]. The coastline of the Atlantic Ocean and the Mediterranean Sea and its size changed accordingly, huge areas, e.g., the North Adriatic Sea, were repeatedly desiccated and reflooded. During the Last Glacial Maximum (24,000–18,000 years BP), the sea level was about 125 m lower than today, with the most intensive sea level rise between 17,000–7000 years BP, after which it reached more or less the present coastline [27]. Glacial and interglacial phases resulted in the sea temperature alterations, with the reconstructed cooling amplitude in the Mediterranean during the Last Glacial Maximum reaching up to 6–7 °C [28], which should have had a severe impact on the diversity and distribution of the living organisms in the Mediterranean Sea. However, the Mediterranean waters remained warmer that those of the adjacent Atlantic Ocean during glacial peaks, thus many species now present in the warm temperate Atlantic likely survived the cold phases of the glacial cycles in the Mediterranean Sea, recolonizing the Atlantic when more favourable temperatures were re-established during interglacial phases like the present one [29].

The studied area concerns the Mediterranean Sea province and the South European Atlantic Shelf ecoregion. The Mediterranean Sea is classified as a specific, well defined biogeographic province within the Temperate Northern Atlantic realm [30]. It is further divided into seven ecoregions (Adriatic Sea, Aegean Sea, Levantine Sea, Tunisian Plateau/Gulf of Sidra, Ionian Sea, Western Mediterranean, and Alboran Sea). Its neighbouring marine provinces are the Black Sea in the east and the Lusitanian province in the Atlantic Ocean in the west. The Mediterranean Sea is connected with the South European Atlantic Shelf and Saharan Upwelling ecoregions (both belonging to the Lusitanian province and having the boundary in the Gibraltar Strait area).

Circulation of water in the studied area, and especially in the Mediterranean Sea, is very complex. The main currents and gyres are depicted in Figure 1. Within the studied region, multiple biogeographical barriers have been identified, of which the Almeria-Oran front, the Strait of Sicily and the Otranto Strait (see Figure 1) are considered to be the major ones influencing genetic diversity of various marine organisms [1,13]. However, dissimilar influence of biogeographic barriers has been found even for the closely related taxa with the same biology and ecology [31–34].

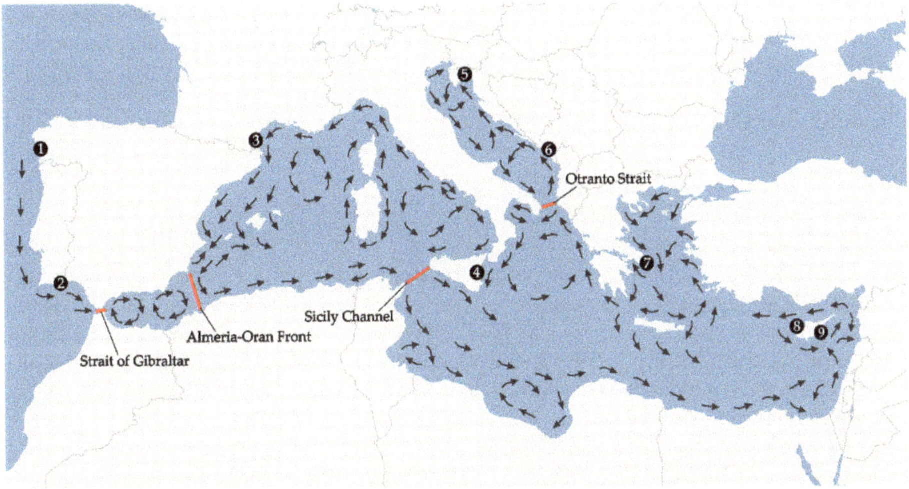

Figure 1. Map of the studied area, with sampling sites (black dots), main oceanographic barriers (red lines) and circulation of the main currents in the Mediterranean Sea (arrows). 1—Spain, 2—Portugal, 3—France, 4—Sicily, 5—Croatia, 6—Montenegro, 7—Greece, 8—Cyprus W (west), 9—Cyprus E (east).

Gobies (Gobiiformes: Gobiidae), with currently recognized 1915 valid species, are the most species-rich family of fishes in the world [35]. It is also the most speciose family of fishes of the European seas and, in particular, of the Mediterranean Sea. The majority of gobies are benthic fishes with planktonic larval stages. Marine gobies predominantly occupy shallow shelf bottoms, with limited number of species extending to deeper shelf, and only a few of them reaching bathyal depths [36,37]. In total, there are more than 90 marine species of gobies in European seas listed in the last review [38]. However, new species are still being discovered, e.g., [36,37,39,40], and the knowledge about the distribution of many species is still quite limited e.g., [41–43]. The information about the population genetic diversity and phylogeography of European marine gobies is very poor too. Population genetic studies have been done so far in only a few species, which are easy to collect by the common fishery techniques. Nevertheless, on the basis of limited information, it is evident that for some studied goby species there is a clear genetic differentiation between distant populations [44–50], but not for the others [51]. Within the genus *Gobius*, the population genetic structure was studied so far in the only species, *Gobius niger*, where the mitochondrial marker nicotinamide adenine dinucleotide dehydrogenase (NADH) was analysed by restriction fragment length polymorphism (RFLP) [51]. The results suggested the existence of population subdivision in this species.

In this work we studied the population genetic diversity of *Gobius geniporus* and *Gobius cruentatus* (*Gobius*-lineage, gobiine-like clade *sensu* Agorreta et al. [52]), two bottom dwelling (epibenthic) species. *Gobius cruentatus* occurs in the north-eastern Atlantic Ocean, from south-western Ireland to the coasts of Senegal, and in the Mediterranean and the Black Seas [53,54], while *G. geniporus* is a Mediterranean endemic [53]. *Gobius cruentatus* is typically detected on mixed bottom habitats dominated by stones, boulders or seagrass [55]. It can grow up to 18 cm and occurs in depths between 1.5 and 40 m [53,54]. *Gobius geniporus* prefers sandy bottoms mixed with gravel, cobbles and boulders, with at least some amount of rocky formations present scattered over sand. It reaches a size of 16 cm and the depth at which it was observed ranges from 1 to 30 m [53,56,57]. These species are restricted to shallow shelf bottoms, so their real area of occupancy is only a narrow stripe of bottom along the coastline. Being small, epibenthic, territorial and non-migratory in adulthood, these species are expected to have only a limited dispersal capability [53].

The aim of this study was to assess the genetic diversity of seven geographically distant populations of two species of European marine gobies, *G. geniporus* and *G. cruentatus*, across their distribution ranges, using mitochondrial and nuclear markers, in order to reveal a possible population subdivision and a potential existence of biogeographical barriers which would affect the connectivity of the populations.

2. Materials and Methods

2.1. Samples

It is very difficult to collect benthic marine gobies in general, unless they live on sandy or muddy bottoms. They are usually not commercially used, so it is not possible to purchase them or to obtain them from fishermen even as a bycatch. As both species investigated in this study typically occupy mixed bottom habitats it is not possible to use common fishery techniques, such as trawl nets or push-nets, to collect them, even though these two species belong to the largest ones among gobies from the north-eastern Atlantic and the Mediterranean Sea. For research purposes, these two species are being collected individually, during scuba-diving and using hand nets and anaesthetic, which is very time consuming. They do not occur in shoals and it can be difficult to spot them. A total of 74 specimens of *G. geniporus* from seven localities in the Mediterranean Sea and 41 specimens of *G. cruentatus* from two localities in the Atlantic Ocean and five localities in the Mediterranean Sea were included in this study (see Figure 1 and Table 1). Tissue samples and voucher specimens are deposited in the ichthyological collection of the National Museum, Prague, Czech Republic.

Table 1. Sampling sites and number of analysed specimens of *Gobius geniporus* and *G. cruentatus* for cytochrome b and S7.

Locality	Coordinates	*G. geniporus*		*G. cruentatus*	
		N cyt b	N S7	N cyt b	N S7
Cyprus E (east)—Cavo Greco	34.98556° N, 34.07667° E	11	11		
Cyprus W (west)—Akamas	35.07528° N, 32.33278° E	12	12	3	3
Greece—Evia Island	37.99694° N, 24.39806° E	10	10		
Montenegro—Boka Kotorska	42.48500° N, 18.67028° E	13	13	2	2
Croatia—Selce	45.15194° N, 14.72083° E	10	10	11	11
Sicily—near Catania	37.99694° N, 24.39806° E	16	15	12	12
France—Banyuls sur Mer	42.48194° N, 3.13667° E	2	2	4	4
Spain—Galicia, Vigo	42.24917° N, 8.75583° W			5	3
Portugal—Algarve	37.07389° N, 8.30361° W			4	4
Total		74	73	41	39

2.2. DNA Extraction, Amplification, Sequencing

Total genomic DNA was extracted from the finclips using Geneaid® DNA Isolation Kit following the manufacturer's protocol. The samples were analysed for two genes, one mitochondrial, cytochrome b (cyt b), and one nuclear, first intron of the ribosomal protein gene S7 (S7). For the amplification of cyt b, the primers GluF and ThrR [58] were used. S7 was amplified with the primers S7RPEX1F and S7RPEX2R [59]. The polymerase chain reaction (PCR) was performed in 25 μL total volume containing 12.5 μL of PPP Master Mix (TopBio), 9.7 μL of Ultrapure H_2O, 0.65 μL of each primer and 2 μL of DNA isolate. Amplification of cyt b followed the protocol described in Šanda et al. [60]. For S7, a specific touch-down protocol was used with the following steps: initial denaturation at 95 °C for 5 min, followed by 5 cycles of denaturation, annealing, and elongation: 94 °C for 40 s, 60 °C for 1 min, 72 °C for 2 min, followed by 35 cycles of denaturation, annealing, and elongation: 95 °C for 30 s, 56 °C for 1 min, 72 °C for 2 min, and the final elongation at 72 °C for 20 min. PCR products were purified with the use of ExoSAP-IT and sequenced at Macrogen Europe. For the sequencing of cyt b, the specific internal primers were designed: GcruF1 (5′-GGT GCA ACC GTC ATC ACT AA-3′) and GcruR1 (5′-AGT GGG TTG GCA GGA ATG-3′) for *G. cruentatus* and GgenF1 (5′-GTA GGC TAT GTC CTG CCC TG AG-3′) and GgenR1 (5′-TTG GAG CCT GTC TCG TG GA-3′) for *G. geniporus*. Nuclear gene S7 was sequenced using the amplification primers. Sequences were deposited in GenBank under accession numbers MT774412-MT774485 (cyt b) and MT893746—MT893891 (S7) for *G. geniporus* and MT684467—MT684507 (cyt b) and MT684508—MT684585 (S7) for *G. cruentatus*.

2.3. Data Analyses

Obtained cyt b and S7 sequences were checked manually in Chromas v2.6.4 and aligned in Bioedit v7.2.6.1 [61]. The appropriate model of nucleotide substitution was determined using jModelTest v2.1.9 [62], based on Akaike Information Criterion (AIC) [63]. DnaSP v6.11.01 [64] was used to assess the haplotype diversity (Hd) and nucleotide diversity (π), as well as to perform Fu and Li's F and Tajima's D neutrality tests. The results of these tests can point to possible selection or a change in population demography. To evaluate the amount of genetic variance within and between populations, analysis of molecular variance (AMOVA) was performed using ARLEQUIN v3.5 [65]. It estimates population differentiation with the use of individual haplotypes and their frequency in the studied populations. This further enables calculation of fixation indices, a global F_{ST} and pairwise F_{ST}s. F_{ST} expresses a degree of genetic differentiation between the individual populations. The two populations from Cyprus were grouped together, as well as the two populations from the Adriatic Sea (Montenegrin and Croatian); the grouping was made based on location, proximity and the water circulation. The remaining populations represented individual groups. The statistical significance of the F_{ST} values was tested by executing 16,000 permutations. For pairwise F_{ST}s, a Bonferroni correction was subsequently applied to correct for multiple tests. Further, genetic distances (uncorrected p-distances) between and within populations of each species were calculated in MEGA 6 [66]. The datasets of S7 were phased by the program PHASE v2.1.1 [67]. All sequences were phased with a probability of 0.9 and the final datasets with inferred phased sequences consisted of 146 sequences for *G. geniporus* and 78 for *G. cruentatus*. The phased S7 data were then used for calculating diversity measures and constructing haplotype networks. The rest of the analyses were not performed on S7 due to a very low polymorphism of S7 datasets. A detailed reconstruction of relationships of the haplotypes of populations was performed by a statistical parsimony method under a 95% connection limit [68], using PopART [69].

Isolation by distance hypothesis was tested by Mantel test [70] using R v3.5 software (package adegenet), executing 1000 permutations. Mantel test compares genetic distances estimated by pairwise F_{ST}s with geographical distances between locations. The matrices of geographical distances were derived from the coordinates of the individual localities. Another approach was applied using ARLEQUIN v3.5 [65], where the shortest marine paths between each pair of localities, estimated from the GoogleEarth, were used in matrices of geographical distances. Statistical significance of the Mantel test was estimated by executing 1000 permutations.

We estimated migration routes between pairs of populations using Migrate-n software v4.4. [71]. Five demographic scenarios were tested for each population pair: (1) model assuming full migration between populations, (2) model assuming migration from the population A to the population B, (3) model assuming migration from B to A, (4) model allowing divergence, where A splits off from B, with migration from B to A after the split, (5) model allowing divergence, where B splits off from A, with migration from A to B after the split. For each scenario, the migration rate and population sizes were estimated; for the scenarios (4) and (5) also the time of divergence between the two populations. Pairs of populations are listed in the Table S1. Migrate-n analyses were conducted using a static heating strategy with four short chains with temperature values of 1.0, 1.5, 3.0, and 1.0×10^6 and a single long chain. 1,000,000 steps were recorded every 100 generations with 200,000 steps discarded as burn-in to ensure the convergence of the analyses. Appropriate mutation model was assessed using jModelTest [62] resulting in Hasegawa-Kishino-Yano (HKY). Priors were set as follows: Bayes-priors = THETA * * UNIFORMPRIOR: 0.001, 0.000 0.0100, Bayes-priors = MIG * * UNIFORMPRIOR: 0.000, 100000.000, 10000.000.

Past population demography of each species was inferred using the linear Bayesian skyline plot model [72], implemented in BEAST v1.8.4 [73]. It allows observing fluctuations of effective population sizes from the present, backwards in time, to the coalescence in the most recent common ancestor, and is expressed graphically. Analyses were conducted under the Bayesian coalescent method, with corresponding nucleotide substitution model for each species and using a strict molecular clock. The x-axis of the plot shows the time in mutation units per nucleotide position and y-axis scaled effective population size. Simulations ran for 100 million Markov chain Monte Carlo (MCMC) steps with sampling every 10,000th generation. Results from three independent runs were combined using LogCombiner and burn-in was set to 20 million iterations in each run. Finally, TRACER v1.7.0 [74] was used to check the parameter estimates and visualize Bayesian skyline plots.

3. Results

3.1. *Gobius geniporus*

In *G. geniporus*, 74 specimens were analysed (Table 1). The alignment of cyt b had a length of 1113 bp and contained 56 polymorphic sites, while there were only two variable sites in the 594 bp long alignment of 146 sequences of S7 (Table 2). A total of 45 haplotypes were found for cyt b and only three haplotypes for S7 within seven Mediterranean populations. The best-fit substitution model selected for cyt b was general time reversible with proportion of invariable sites (GTR+I). Haplotype diversity of cyt b was high, while nucleotide diversity low (Hd = 0.969; π = 0.004) and for S7 both haplotype and nucleotide diversity were extremely low (Hd = 0.054; π = 0.0001) (Table 2). Diversity measures calculated per each locality are listed in Table S2. The values of neutrality tests (Tajima's D, Fu and Li's F) for cyt b were negative and significant, indicating a recent population expansion or purifying selection (Table 2). The Bayesian skyline plot of *G. geniporus* depicts a gradual population size growth since the coalescence and its stabilization in the present (Figure 2a).

Table 2. Diversity measures and results of neutrality tests for *Gobius geniporus* and *G. cruentatus* based on cytochrome b and S7 sequences. N—number of sequences, S—number of segregating sites, Nh—number of haplotypes, Hd—haplotype diversity, π—nucleotide diversity. Significant values (at $\alpha = 0.05$) indicated by asterisk.

Species	N	S	Nh	Hd	π	Fu & Li's F	Tajima's D
G. geniporus							
cyt b	74	56	45	0.969	0.004	−3.987 *	−2.055 *
S7	146	2	3	0.054	0.0001		
G. cruentatus							
cyt b	41	47	32	0.985	0.006	−2.522 *	−1.284
S7	78	3	4	0.212	0.0004		

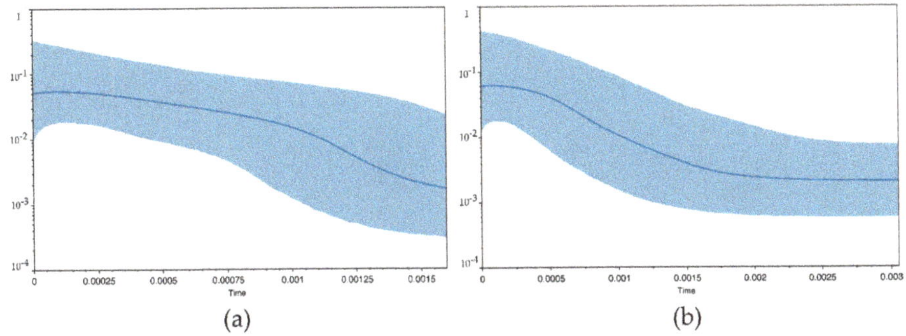

Figure 2. Bayesian skyline plot for *Gobius geniporus* (**a**) and *G. cruentatus* (**b**) based on cytochrome b sequence data. The graph illustrates fluctuation of the population size from recent to the coalescence (from left to right). Y-axis stands for scaled effective population size and x-axis for time scale in units of mutation per nucleotide position. Blue middle line shows the median estimate, 95% confidence interval is indicated in blue.

In the haplotype network of *G. geniporus* based on cyt b, there was an indication of a certain geographical pattern: there were two major haplotype groups. In one of them there was one more frequent haplotype shared between the central Mediterranean populations (Montenegrin, Croatian and Sicilian), while in the other one, there were two more frequent haplotypes, shared between Cypriot and Greek populations (Figure 3a and Figure S1a). Most haplotypes from Cypriot populations grouped together, as well as the majority of the haplotypes from the Sicilian one. Unique haplotypes prevailed in the network. Practically all sequences of S7 in *G. geniporus* were of the same haplotype (142 out of 146, Figure 3a) and due to this low polymorphism, all following analyses were performed only on the cyt b dataset. AMOVA for *G. geniporus* based on the cyt b showed that most of the genetic variance is distributed within populations (76%). F_{ST} index indicated a high level of genetic differentiation ($F_{ST} = 0.237$, $p < 0.01$). Pairwise F_{ST}s showed in most cases a pronounced or high level of genetic differentiation between the pairs of populations, but several values were low (Table 3); however, only a half of values were significant. Statistically significant values indicating high or pronounced differentiation were for most comparisons for Sicilian and both Cypriot populations. Mean p-distances between the populations were low and ranged between 0.2 and 0.6% (Table 3). Mean p-distances within populations were of a similar range (0.1–0.5%), while the maximum intraspecific p-distance for *G. geniporus* was 1.08%.

Table 3. Mean genetic distances between *Gobius geniporus* populations for cytochrome b (uncorrected p-distances, in %, above the diagonal), intrapopulation distances (on diagonal), and pairwise F_{ST}s (below diagonal). Significant values of F_{ST}s (at $\alpha = 0.05$/number of pairs) indicated by asterisk.

	Sicily	Croatia	Montenegro	France	Cyprus W	Cyprus E	Greece
Sicily	0.4	0.5	0.5	0.3	0.4	0.4	0.5
Croatia	0.224 *	0.5	0.4	0.5	0.5	0.6	0.5
Montenegro	0.209 *	0.000	0.4	0.4	0.4	0.5	0.4
France	0.000	0.227	0.263	0.1	0.3	0.3	0.4
Cyprus W	0.285 *	0.369 *	0.349 *	0.408	0.2	0.2	0.3
Cyprus E	0.271 *	0.413 *	0.399 *	0.373	0.056	0.2	0.3
Greece	0.151 *	0.106	0.065	0.160	0.077	0.145 *	0.4

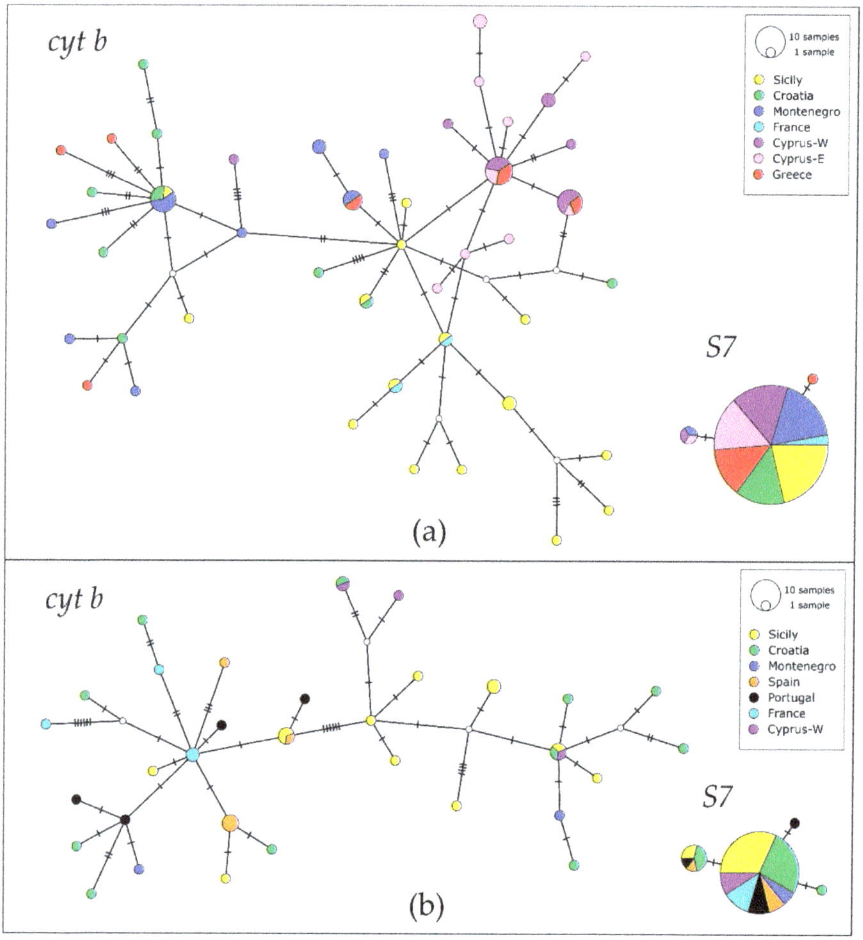

Figure 3. Haplotype networks of *Gobius geniporus* (**a**) and *G. cruentatus* (**b**) based on cytochrome b and S7 sequences. Size of circle is proportional to haplotype frequency. The number of mutational steps between closest haplotypes is indicated by hatch marks. Missing intermediate haplotypes are shown as small white circles.

Mantel test was significant using both approaches (adegenet: observation 0.299, expectation 10^{-5}, p-value < 0.001; ARLEQUIN: correlation coefficient 0.69, p-value < 0.001), indicating a possible pattern of isolation by distance (see Figure 4a). Among the modelled migration scenarios, for each pair of populations, the model which allows divergence and the recurrent immigration from the ancestral population after the split was the one with the highest probability. The divergence directions and the migration routes are schematically depicted in the Figure 5a, while the estimates of immigration rates, divergence times and population sizes are listed in the Table S3. The system of migration routes is rather circular, anticlockwise, with a large circle between Sicily, western Cyprus, eastern Cyprus, Greece, Montenegro, Croatia and Sicily, and two smaller ones: Sicily, Greece, Montenegro, Croatia, Sicily, and Sicily, Montenegro, Croatia, Sicily. All the routes, with the exception of the one between Sicily and Greece eastwards, can be well explained by the prevailing currents (see Figure 5a). The highest rate of migration among the modelled pairs of populations was estimated between the western and eastern Cyprus, correspondingly with their proximity and the prevailing eastward current.

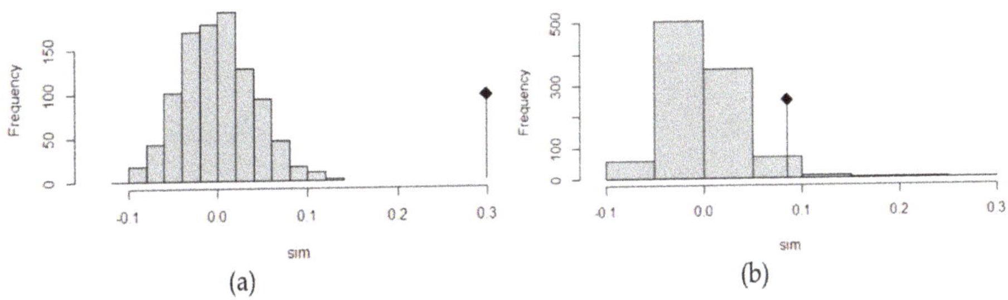

Figure 4. Histograms of the Mantel test assessing the relationship between genetic and geographic distance for *Gobius geniporus* (**a**) and *G. cruentatus* (**b**).

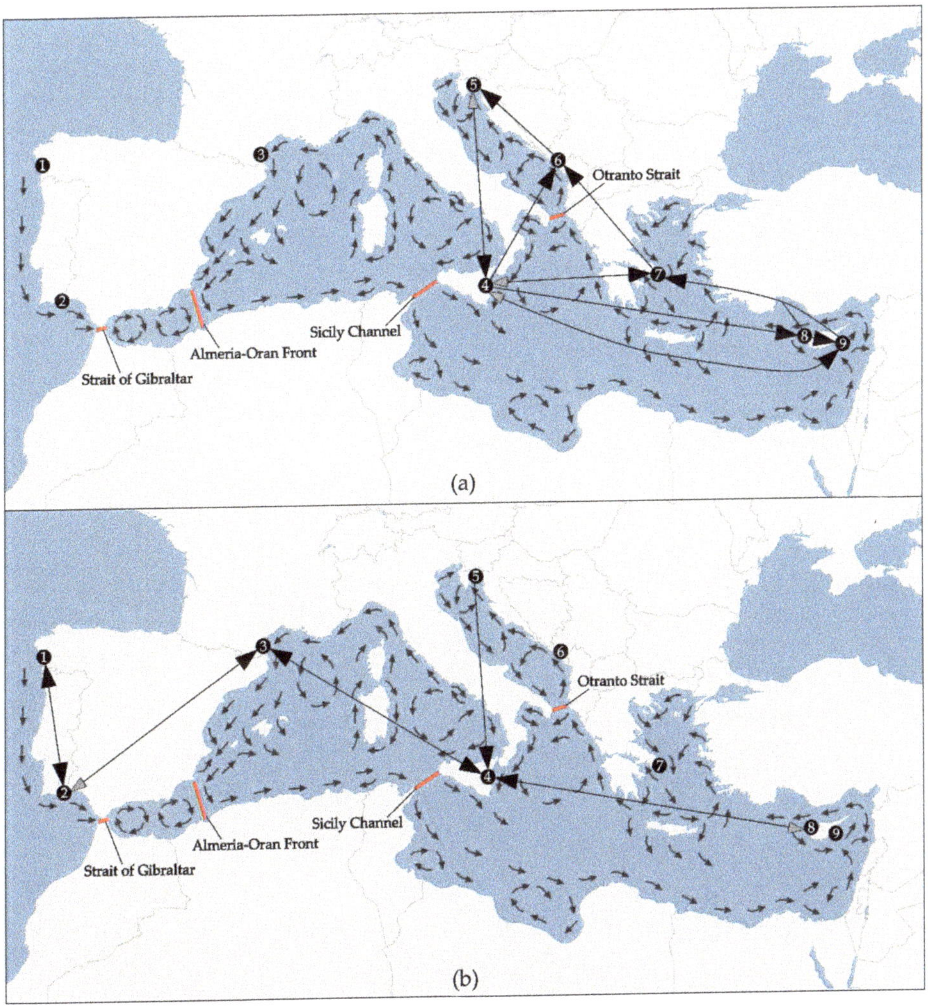

Figure 5. Migration routes of *Gobius geniporus* (**a**) and *G. cruentatus* (**b**) estimated by Migrate-n. Grey tips of arrows indicate less probable directions.

3.2. *Gobius cruentatus*

Mitochondrial marker cyt b was analysed in 41 specimens of *G. cruentatus*, and nuclear marker S7 in 39 specimens (Table 1). The alignment length of cyt b was 1117 bp and contained 47 polymorphic sites, while that of S7 had a length of 555 bp and contained only three segregating sites (Table 2). There were 32 haplotypes of cyt b and only four of S7. The best-fit substitution model selected for cyt b was general time reversible with proportion of invariable sites (GTR + I). An overall haplotype diversity of cyt b was high, while nucleotide diversity low (Hd = 0.985; π = 0.006), which can indicate a recent population expansion. This was also suggested by negative values of neutrality tests (see Table 2). On the other hand, for nuclear gene S7, values of haplotype and nucleotide diversity were markedly lower (Hd = 0.212; π = 0.0004). Diversity measures calculated per each locality are listed in Table S2. The Bayesian skyline plot showed a constant population size in the past, followed by a gradual population expansion and a stable population in the present (Figure 2b).

In *G. cruentatus*, cyt b haplotype network reconstruction did not reveal any well-defined spatial structure (Figure 3b). The network consisted mostly of unique haplotypes, very diverse for each locality (Figure S1b). Three haplotypes were shared between two or three distant populations (Spain/Sicily, Sicily/Croatia/Cyprus, Cyprus/Croatia). The network displayed two highly variable haplogroups separated by six mutational steps. Interestingly, there was a particular geographic pattern: while haplotypes from the central Mediterranean populations (Croatia, Sicily and Montenegro) were in both haplogroups, the haplotypes from the westernmost populations (Portugal, Spain and France) were placed only in one haplogroup, and haplotypes from the easternmost population, Cyprus, occurred only in the other haplogroup. On the contrary, there was nearly no polymorphism in S7. The network was formed by one dominant haplotype which included 69 alleles (out of 78) and was shared among all populations, one less frequent shared haplotype, which included only 7 alleles, and two unique ones. All haplotypes were very similar (Figure 3b). The remaining analyses were performed only on the cyt b dataset due to the low polymorphism in S7.

Similarly to the results of haplotype network, AMOVA performed on the cyt b showed that most of the genetic variance is distributed within the populations, with a ratio of approximately 1:4 between the variability among and within populations. Computed F_{ST} index indicated a high level of genetic differentiation ($F_{ST} = 0.216$, $p < 0.01$). Some values of pairwise F_{STs} performed on *G. cruentatus* indicated a pronounced differentiation (the highest values were between the easternmost population, Cyprus, and three westernmost populations—Spain, Portugal and France), but the majority were low or moderate; however, none of the values was significant (Table 4). The mean p-distances between the populations ranged between 0.3 and 1%, and the range of intrapopulation p-distances was similar (0.2–1.1%) (Table 4). The highest interpopulation divergences were between the westernmost and easternmost populations (0.9–1%, see Table 4). The highest overall intraspecific p-distance for *G. cruentatus* was 1.52%.

Table 4. Mean genetic distances between *Gobius cruentatus* populations for cytochrome b (uncorrected p-distances, in %, above diagonal), intrapopulation distances (on diagonal), and pairwise F_{STs} (below diagonal). None of the values of F_{STs} was significant at α = 0.05/number of pairs.

	Sicily	Croatia	Montenegro	Spain	Portugal	France	Cyprus W
Sicily	0.5	0.7	0.6	0.7	0.7	0.7	0.5
Croatia	0.000	0.8	0.7	0.7	0.7	0.8	0.7
Montenegro	0.000	0.000	1.1	0.7	0.6	0.7	0.7
Spain	0.370	0.223	0.309	0.2	0.3	0.3	0.9
Portugal	0.342	0.176	0.149	0.137	0.2	0.3	0.9
France	0.332	0.172	0.126	0.103	0.044	0.4	1.0
Cyprus W	0.074	0.057	0.077	0.725	0.688	0.613	0.4

The result of the Mantel test was significant using both approaches (adegenet: observation 0.085, expectation 0.002, *p*-value 0.03; ARLEQUIN: correlation coefficient 0.81, *p*-value < 0.01), indicating a possible pattern of isolation by distance (see Figure 4b). According to our results, the most probable scenario of migration of *G. cruentatus* between the studied localities was also the one allowing divergence and the migration from the ancestral population after the split. The divergence and migration routes are schematically depicted in the Figure 5b, while the estimates of immigration rates, divergence times and population sizes are listed in the Table S3. The results were not unequivocal, as can be seen from the Figure 5b, in several cases models proposing the opposite direction of divergence and migration between two populations had almost the same probability. This might be due to the low number of samples.

4. Discussion

Many previous phylogeographic studies have shown the existence of the geographical structure in populations of zoobiota within the north-eastern Atlantic and Mediterranean (i.e., Northern European Seas, Lusitanian and Mediterranean Sea provinces of Temperate Northern Atlantic realm *sensu* Spalding et al. [30]), e.g., [7,8,10–12]. In this term, however, the European marine gobies, despite being the most speciose and abundant fish family in this area, have been little studied so far.

Our results on two epibenthic goby species (*G. geniporus* and *G. cruentatus*) showed that the most plausible model which can explain the genetic structure of populations of both species is a model of divergence and recurrent migration from the ancestral population after the split. In the case of *G. geniporus*, the direction of divergence and the migration routes match well the prevailing currents between the studied localities (Figure 5a). The main feature of the migration route is anticlockwise circulation from Sicily towards east and then turning westwardly back to Sicily, and making a smaller circle from Sicily to the Adriatic Sea, following the Montenegrin and Croatian coast and subsequently the Italian coast, and back to Sicily. More dense sampling would be useful to confirm these findings, as also smaller gyres can substantially influence the genetic structure of epibenthic fishes [75]. The directions of divergence and migration between the pairs of populations of *G. cruentatus* were ambiguous. The observed pattern may be an outcome of a low number of individuals used to infer the migration routes in this species. Alternatively, it might be a consequence of higher complexity of water circulation in the species range, and/or biology of this species (see later discussion on hyperbenthic juveniles).

The lifestyle of the two studied species matches the model of divergence and recurrent migration. Being epibenthic and territorial, *G. geniporus* and *G. cruentatus* most probably exhibit only short-distance movements in adulthood, which allow the divergence between populations. Their main dispersal route is thus via a transport of planktonic larval stages, which can be dispersed by currents. The distance which a larva can reach depends mainly on the hydrodynamics and on the duration and behaviour of the larval stage, but the dispersal of planktonic larvae is much more complex and still not well understood [76]. The high multiscale variability of topography, temperature and salinity in the Mediterranean Sea generates free and boundary currents, bifurcating jets, meander and ring vortices, permanent or temporary cyclonic and anticyclonic gyres and eddies [77]. Recently, computer simulations that integrate a high number of biological and marine physical information have been successfully used in several works focused on the role of marine currents on the dispersion and genetic structure of marine organisms [75,78–80].

The influence of currents on genetic structure of the populations of epibenthic marine fish species was found for *Tripterygion tripteronotum* [75], where the population structure matched well the gyres in the Adriatic Sea, and also for other marine organisms [78–82].

Where known, the planktonic life stages in different European goby species have a variable duration, with a minimum of 13 days in *Zosterisessor ophiocephalus* to 51 days in *Gobius paganellus* [83]. However, in many Mediterranean gobies, nothing is known about their larvae, the duration of this stage, nor about their dispersion routes or distances. A similar range of planktonic larval duration (PLD) was observed in other Mediterranean fish species. In epibenthic Mediterranean littoral fish species of the genus *Tripterygion*, the PLD is estimated to be two to three weeks [84], while blennies

(Blenniidae) have a PLD between 22 and 71 days. In species of both these fish groups a population genetic subdivision was observed [75,85–87].

The dispersal capability of fish larvae can broadly differ, while it is around 120 km during 80–170 days of PLD in *Sebastes melanops* [76,88], it is only 100–500 m during 30–50 days in *Chaetodon vagabundus* [76,89]. This underlines the complexity of the dispersion process of fish larval stages.

Apart from having mobile larval stages, *G. cruentatus* has hyperbenthic juveniles, swimming in shoals within 1 m above the sea bottom. It is not known which distances this stage can cover and whether the dispersion during this stage has any influence on the gene flow. In the aquarium, this stage lasted two months [90]. It is not known whether other European gobies have hyperbenthic juveniles. Also other biological traits, such as reproduction strategy (European benthic gobies are iteroparous [53]) and timing of spawning, can influence the genetic structure of populations [75].

Our results showed that there was a high diversity of haplotypes of cyt b at each sampled locality. As discussed above, no clear population subdivision was found in two studied species, as it was disturbed by the recurrent migrations between the populations. There was a certain structuring in both species, as two haplogroups are observable in the networks (Figure 3). In *G. geniporus*, in the most frequent haplotypes of each haplogroup, specimens from different areas dominate: in one, the specimens from the eastern (Cyprus and Greece), while in the other, the specimens from the central Mediterranean Sea (Italy, Montenegro and Croatia). However, haplotypes of specimens from the Sicilian population are prevailing in the haplogroup with the eastern Mediterranean Sea haplotypes. In *G. cruentatus*, one haplogroup includes all specimens from the western part of the species range, from the Atlantic coast of Spain and Portugal, as well as from the western Mediterranean French coast, while the other haplogroup includes all samples from the eastern Mediterranean Sea (Cyprus). However, the central Mediterranean samples (Sicily, Montenegro and Croatia) are present in both haplogroups. Similar situation, where haplotypes from different haplogroups were found at the same geographic locality, with no clear geographical pattern, was observed also in other fish species in the Mediterranean Sea [91,92]. It was attributed to the secondary contact between the isolated populations which diverged in allopatry and came to a contact again after the removal of the migration barrier [91]. Additionally, our migration scheme for *G. cruentatus* shows convergence of the routes from the eastern and western Mediterranean Sea and the Adriatic Sea near Sicily, corresponding to the situation in the haplotype network.

Most of the research on population genetic structure of marine gobies from Europe have been conducted on epibenthic species of the genus *Pomatoschistus* (gobionelline-like gobies [52]), usually inhabiting lagoons and shallow coastal waters with fine substrates. Population genetic differentiation was observed in all four studied *Pomatoschistus* species [44–50,93]. Population genetic diversity of species from the gobiine-like gobies [52] has been studied in only two European marine species [51], epibenthic goby *G. niger*, living on the muddy substrates, and *Aphia minuta*, a pelagic shoal species. Giovannotti et al. [51] found a spatial genetic structure in epibenthic *G. niger*, while no structure in the pelagic *A. minuta*.

There are several recognised biogeographic breaks in the Mediterranean Sea and the north-eastern Atlantic Ocean. Our data did not point to the existence of any biogeographic boundary preventing a gene flow between the studied populations for neither of the two species. However, the effect of a small sample size cannot be excluded. The Strait of Gibraltar, or rather the Almeria-Oran front, which is an important biogeographic barrier for some marine organisms [4,8,10,33,94], did not have any influence on the gene flow between Atlantic and western Mediterranean populations of *G. cruentatus*. Similarly, this break does not present a barrier to gene flow of the various fish species, neither pelagic, e.g., *Sardina pilchardus* (nDNA microsatellite loci) [95], *Thunnus thynnus* (mtDNA d-loop) [34], *Scomber colias* (mtDNA d-loop) [31], *Diplodus sargus* (mtDNA d-loop, nDNA S7 first intron) [96], nor benthic ones, ranging from widespread eurybathic *Lophius piscatorius* (mtDNA d-loop), able to reach depths down to 500 m [33] to *Parablennius sanguinolentus* (mtDNA d-loop, nDNA S7 first intron), which is restricted to very shallow littoral of 0–1 m depth [86,87]. Neither did the Sicily Channel influence the genetic structure of the two studied goby species, unlike is the case of some other fish species, e.g., *Dicentrarchus labrax* (nDNA microsatellite loci) [97], *Sprattus sprattus* (mtDNA d-loop) [11], and *P. tortonesei* (mtDNA 16S, COI) [48], where the Sicily Channel presents an important breakpoint. Although many studies showed a genetic differentiation between populations of the biota of the Adriatic and the Mediterranean Seas, separated by the Otranto strait, e.g., in *P. minutus* (mtDNA d-loop, cyt b, allozymes) [44–46,50], *Platichthys flesus* (allozymes) [98], Gouania willdenowi (mtDNA COI and 9 nDNA markers) [99] and *Sparus aurata* (allozymes) [100], neither was the Otranto Strait a biogeographic barrier for *G. cruentatus* and *G. geniporus*.

5. Conclusions

Our data revealed that the population genetic structure of the two studied epibenthic goby species (*G. geniporus* and *G. cruentatus*) can be well explained by the model of migration, allowing divergence between each pair of populations, with the ongoing migration from the ancestral population. This corresponds well with the biology of these gobies, having poorly mobile adults on one hand, and planktonic larval stages, which can be dispersed by currents, on the other hand. The population genetic structure of *G. geniporus* is influenced by currents: the estimated migration routes between the studied populations follow the main current directions in the study area.

Supplementary Materials: The following are available online at http://www.mdpi.com/2077-1312/8/10/792/s1, Figure S1: List of pairs of populations modelled in Migrate-n, Table S2: Diversity measures for *Gobius geniporus* and *G. cruentatus* calculated per each locality based on cytochrome b and S7 sequences, Table S3: Posterior distribution table of Migrate-n analyses, Figure S1: Cytochrome b haplotype frequencies at each locality, *Gobius geniporus* (a), *G. cruentatus* (b).

Author Contributions: Conceptualization, J.V. and R.Š.; methodology, K.Č., J.V., K.E., R.Š.; software, K.Č., K.E., J.V.; validation, J.V., A.M.P.; formal analysis, K.Č., J.V., K.E.; investigation, K.Č., J.V., R.Š., T.S.; resources, J.V., R.Š., M.K., S.Z., A.M.P.; data curation, K.Č.; writing—original draft preparation, K.Č.; writing—review and editing, J.V., R.Š., M.K., K.E., S.Z., A.M.P., T.S.; visualization, K.Č., J.V.; supervision, J.V.; project administration, J.V.; funding acquisition, J.V., R.Š., T.S., K.Č. All authors have read and agreed to the published version of the manuscript.

Funding: This research was funded by the Grant Agency of the Charles University (GAUK), grant number 1192217. Stay of K.Č. at the University of Catania was supported by Erasmus program. RŠ received support by the Ministry of Culture of the Czech Republic (DKRVO 2019–2023/6.III.b National Museum, 00023272).

Acknowledgments: We are very grateful to two anonymous reviewers for their constructive comments, which helped us to improve our manuscript. We wish to thank David Villegas Rios (Spain), Konstantinos Moustakas and Ioulianos Pantelides (DFMR, Cyprus) and the Dikelas Dive Center (Karystos, Greece) for the help in the field. Sampling in Greece and Cyprus was conducted under permission of the Hellenic Ministry of Environment (through collecting licence to HCMR, no. 220965/2583/22-8-2011) and the Cyprus Ministry of Agriculture, Rural Development and Environment, through the Department of Fisheries and Marine Research (no. ΠΠ2921ΠΠ/10-11-2017) respectively. Collection of a part of the material was supported by the EU FP7 project ASSEMBLE at CCMar/Centre of Marine Sciences of Algarve, Faro, Portugal, and Observatoire oceanologique de Banyuls/Mer, Laboratoire Arago, Banyuls/Mer.

Conflicts of Interest: The authors declare no conflict of interest.

References

1. Pascual, M.; Rives, B.; Schunter, C.; Macpherson, E. Impact of life history traits on gene flow: A multispecies systematic review across oceanographic barriers in the Mediterranean Sea. *PLoS ONE* **2017**, *12*, e0176419. [CrossRef] [PubMed]
2. Coyer, J.A.; Diekmann, O.E.; Serrão, E.A.; Procaccini, G.; Milchakova, N.; Pearson, G.A.; Stam, W.T.; Olsen, J.L. Population genetics of dwarf eelgrass *Zostera noltii* throughout its biogeographic range. *Mar. Ecol. Prog. Ser.* **2004**, *281*, 51–62. [CrossRef]
3. Riesgo, A.; Taboada, S.; Pérez-Portela, R.; Melis, P.; Xavier, J.R.; Blasco, G.; López-Legentil, S. Genetic diversity, connectivity and gene flow along the distribution of the emblematic Atlanto-Mediterranean sponge *Petrosia ficiformis* (Haplosclerida, Demospongiae). *BMC Evol. Biol.* **2019**, *19*, 24. [CrossRef] [PubMed]

4. Sá-Pinto, A.; Branco, M.S.; Alexandrino, P.B.; Fontaine, M.C.; Baird, S.J.E. Barriers to gene flow in the marine environment: Insights from two common intertidal limpet species of the Atlantic and Mediterranean. *PLoS ONE* **2012**, *7*, e50330. [CrossRef]
5. Borrero-Pérez, G.H.; González-Wangüemert, M.; Marcos, C.; Pérez-Ruzafa, A. Phylogeography of the Atlanto-Mediterranean sea cucumber *Holothuria* (*Holothuria*) *mammata*: The combined effects of historical processes and current oceanographical pattern. *Mol. Ecol.* **2011**, *20*, 1964–1975. [CrossRef]
6. Penant, G.; Aurelle, D.; Feral, J.P.; Chenuil, A. Planktonic larvae do not ensure gene flow in the edible sea urchin *Paracentrotus lividus*. *Mar. Ecol. Prog. Ser.* **2013**, *480*, 155–170. [CrossRef]
7. Palero, F.; Abelló, P.; Macpherson, E.; Gristina, M.; Pascual, M. Phylogeography of the European spiny lobster (*Palinurus elephas*): Influence of current oceanographical features and historical processes. *Mol. Phylogenet. Evol.* **2008**, *48*, 708–717. [CrossRef]
8. Reuschel, S.; Cuesta, J.A.; Schubart, C.D. Marine biogeographic boundaries and human introduction along the European coast revealed by phylogeography of the prawn *Palaemon elegans*. *Mol. Phylogenet. Evol.* **2010**, *55*, 765–775. [CrossRef]
9. Liu, H.; Li, S.; Ugolini, A.; Momtazi, F.; Hou, Z. Tethyan closure drove tropical marine biodiversity: Vicariant diversification of intertidal crustaceans. *J. Biogeogr.* **2018**, *45*, 941–951. [CrossRef]
10. Cimmaruta, R.; Bondanelli, P.; Nascetti, G. Genetic structure and environmental heterogeneity in the European hake (*Merluccius merluccius*). *Mol. Ecol.* **2005**, *14*, 2577–2591. [CrossRef]
11. Debes, P.V.; Zachos, F.E.; Hanel, R. Mitochondrial phylogeography of the European sprat (*Sprattus sprattus* L., Clupeidae) reveals isolated climatically vulnerable populations in the Mediterranean Sea and range expansion in the northeast Atlantic. *Mol. Ecol.* **2008**, *17*, 3873–3888. [CrossRef] [PubMed]
12. Magoulas, A.; Castilho, R.; Caetano, S.; Marcato, S.; Patarnello, T. Mitochondrial DNA reveals a mosaic pattern of phylogeographical structure in Atlantic and Mediterranean populations of anchovy (*Engraulis encrasicolus*). *Mol. Phylogenet. Evol.* **2006**, *39*, 734–746. [CrossRef] [PubMed]
13. Patarnello, T.; Volckaert, F.A.M.J.; Castilho, R. Pillars of Hercules: Is the Atlantic-Mediterranean transition a phylogeographical break? *Mol. Ecol.* **2007**, *16*, 4426–4444. [CrossRef] [PubMed]
14. Palumbi, S.R. Genetic divergence, reproductive isolation, and marine speciation. *Annu. Rev. Ecol. Syst.* **1994**, *25*, 547–572. [CrossRef]
15. Palumbi, S.R. Marine reserves and ocean neighborhoods: The spatial scale of marine populations and their management. *Annu. Rev. Environ. Resour.* **2004**, *29*, 31–68. [CrossRef]
16. Thiede, J. A Glacial Mediterranean. *Nature* **1978**, *276*, 680–683. [CrossRef]
17. Roveri, M.; Flecker, R.; Krijgsman, W.; Lofi, J.; Lugli, S.; Manzi, V.; Sierro, F.J.; Bertini, A.; Camerlenghi, A.; De Lange, G.; et al. The Messinian Salinity Crisis: Past and future of a great challenge for marine sciences. *Mar. Geol.* **2014**, *352*, 25–58. [CrossRef]
18. Hsü, K.J.; Ryan, W.B.F.; Cita, M.B. Late miocene desiccation of the Mediterranean. *Nature* **1973**, *242*, 240–244. [CrossRef]
19. Hsü, K.J.; Montadert, L.; Bernoulli, D.; Cita, M.B.; Erickson, A.; Garrison, R.E.; Kidd, R.B.; Mèlierés, F.; Müller, C.; Wright, R. History of the Mediterranean salinity crisis. *Nature* **1977**, *267*, 1053–1078. [CrossRef]
20. Garcia-Castellanos, D.; Villaseñor, A. Messinian salinity crisis regulated by competing tectonics and erosion at the Gibraltar arc. *Nature* **2011**, *480*, 359–363. [CrossRef]
21. Carnevale, G.; Longinelli, A.; Caputo, D.; Barbieri, M.; Landini, W. Did the Mediterranean marine reflooding precede the Mio-Pliocene boundary? Paleontological and geochemical evidence from upper Messinian sequences of Tuscany, Italy. *Palaeogeogr. Palaeoclimatol. Palaeoecol.* **2008**, *257*, 81–105. [CrossRef]
22. Domingues, V.S.; Bucciarelli, G.; Almada, V.C.; Bernardi, G. Historical colonization and demography of the *Mediterranean damselfish, Chromis chromis*. *Mol. Ecol.* **2005**, *14*, 4051–4063. [CrossRef] [PubMed]
23. Carnevale, G.; Landini, W.; Sarti, G. Mare versus Largo-mare: Marine fishes and the Mediterranean environment at the end of the Messinian Salinity Crisis. *J. Geol. Soc. London.* **2006**, *163*, 75–80. [CrossRef]
24. Garcia-Castellanos, D.; Estrada, F.; Jiménez-Munt, I.; Gorini, C.; Fernández, M.; Vergés, J.; De Vicente, R. Catastrophic flood of the Mediterranean after the Messinian salinity crisis. *Nature* **2009**, *462*, 778–781. [CrossRef] [PubMed]
25. Wilson, A.B.; Veraguth, I.E. The impact of Pleistocene glaciation across the range of a widespread European coastal species. *Mol. Ecol.* **2010**, *19*, 4535–4553. [CrossRef] [PubMed]

26. Rohling, E.J.; Grant, K.; Bolshaw, M.; Roberts, A.P.; Siddall, M.; Hemleben, C.; Kucera, M. Antarctic temperature and global sea level closely coupled over the past five glacial cycles. *Nat. Geosci.* **2009**, *2*, 500–504. [CrossRef]
27. Fleming, K.; Johnston, P.; Zwartz, D.; Yokoyama, Y.; Lambeck, K.; Chappell, J. Refining the eustatic sea-level curve since the Last Glacial Maximum using far- and intermediate-field sites. *Earth Planet. Sci. Lett.* **1998**, *163*, 327–342. [CrossRef]
28. Essallami, L.; Sicre, M.A.; Kallel, N.; Labeyrie, L.; Siani, G. Hydrological changes in the Mediterranean Sea over the last 30,000 years. *Geochem. Geophys. Geosyst.* **2007**, *8*. [CrossRef]
29. Almada, V.C.; Oliveira, R.F.; Goncalves, E.J.; Almeida, J.; Santos, R.S.; Wirtz, P. Patterns of diversity of the north-eastern Atlantic blenniid fish fauna (Pisces: Blenniidae). *Glob. Ecol. Biogeogr.* **2001**, *10*, 411–422. [CrossRef]
30. Spalding, M.D.; Fox, H.E.; Allen, G.R.; Davidson, N.; Ferdaña, Z.A.; Finlayson, M.; Halpern, B.S.; Jorge, M.A.; Lombana, A.; Lourie, S.A.; et al. Marine ecoregions of the world: A bioregionalization of coastal and shelf areas. *Bioscience* **2007**, *57*, 573–583. [CrossRef]
31. Zardoya, R.; Castilho, R.; Grande, C.; Favre-Krey, L.; Caetano, S.; Marcato, S.; Krey, G.; Patarnello, T. Differential population structuring of two closely related fish species, the mackerel (*Scomber scombrus*) and the chub mackerel (*Scomber japonicus*), in the Mediterranean Sea. *Mol. Ecol.* **2004**, *13*, 1785–1798. [CrossRef] [PubMed]
32. Bargelloni, L.; Alarcon, J.A.; Alvarez, M.C.; Penzo, E.; Magoulas, A.; Palma, J.; Patarnello, T. The Atlantic-Mediterranean transition: Discordant genetic patterns in two seabream species, *Diplodus puntazzo* (Cetti) and *Diplodus sargus* (L.). *Mol. Phylogenet. Evol.* **2005**, *36*, 523–535. [CrossRef] [PubMed]
33. Charrier, G.; Chenel, T.; Durand, J.D.; Girard, M.; Quiniou, L.; Laroche, J. Discrepancies in phylogeographical patterns of two European anglerfishes (*Lophius budegassa* and *Lophius piscatorius*). *Mol. Phylogenet. Evol.* **2006**, *38*, 742–754. [CrossRef] [PubMed]
34. Alvarado Bremer, J.R.; Viñas, J.; Mejuto, J.; Ely, B.; Pla, C. Comparative phylogeography of Atlantic bluefin tuna and swordfish: The combined effects of vicariance, secondary contact, introgression, and population expansion on the regional phylogenies of two highly migratory pelagic fishes. *Mol. Phylogenet. Evol.* **2005**, *36*, 169–187. [CrossRef] [PubMed]
35. Fricke, R.; Eschmeyer, W.N.; Fong, J.D. Eschemeyer's Catalog of Fishes. Available online: https://www.researchgate.net/project/Eschmeyers-Catalog-of-Fishes-online (accessed on 12 October 2020).
36. Kovačić, M.; Ordines, F.; Schliewen, U.K. A new species of *Buenia* (Perciformes: Gobiidae) from the Western Mediterranean slope bottoms, the redescription of *Buenia jeffreysi* and the first Balearic record of *Buenia affinis*. *Zootaxa* **2018**, *4392*, 267–288. [CrossRef] [PubMed]
37. Kovačić, M.; Ordines, F.; Ramirez-Amaro, S.; Schliewen, U.K. *Gymnesigobius medits* (Teleostei: Gobiidae), a new gobiid genus and species from the western Mediterranean slope bottoms. *Zootaxa* **2019**, *4651*, 513–530. [CrossRef] [PubMed]
38. Patzner, R.A.; Van Tassell, J.L.; Kovačić, M.; Kapoor, B.G. *The Biology of Gobies*; Science Publishers: New York, NY, USA; CRC Press: New York, NY, USA; Taylor & Francis Group: New York, NY, USA, 2011; ISBN 9781578084364.
39. Kovačić, M.; Šanda, R. A new species of *Gobius* (Perciformes: Gobiidae) from the Mediterranean Sea and the redescription of *Gobius bucchichi*. *J. Fish Biol.* **2016**, *88*, 1104–1124. [CrossRef]
40. Engin, S.; Seyhan, D. A new species of *Pomatoschistus* (Teleostei, Gobiidae): The Mediterranean's smallest marine fish. *J. Fish Biol.* **2017**, *91*, 1208–1223. [CrossRef]
41. Kovačić, M.; Šanda, R.; Kirinčić, M.; Zanella, D. Geographic distribution of gobies (Gobiidae) in the Adriatic Sea with thirteen new records for its southern part. *Cybium* **2012**, *36*, 435–445.
42. Engin, S.; Irmak, E.; Seyhan, D.; Akdemir, T.; Keskin, A.C. Gobiid fishes of the coastal zone of the Northeastern Aegean Sea. *Mar. Biodivers.* **2018**, *48*, 1073–1084. [CrossRef]
43. Kovačić, M.; Schembri, P.J. Twelve new records of gobies and clingfishes (Pisces: Teleostei) significantly increase small benthic fish diversity of Maltese waters. *Mediterr. Mar. Sci.* **2019**, *20*, 287–296. [CrossRef]
44. Stefanni, S.; Gysels, E.S.; Volckaert, F.A.M.; Miller, P.J. Allozyme variation and genetic divergence in the sand goby, *Pomatoschistus minutus* (Teleostei: Gobiidae). *J. Mar. Biol. Assoc. UK* **2003**, *83*, 1143–1149. [CrossRef]

45. Stefanni, S.; Thorley, J.L. Mitochondrial DNA phylogeography reveals the existence of an Evolutionarily Significant Unit of the sand goby *Pomatoschistus minutus* in the Adriatic (Eastern Mediterranean). *Mol. Phylogenet. Evol.* **2003**, *28*, 601–609. [CrossRef]
46. Gysels, E.S.; Hellemans, B.; Patarnello, T.; Volckaert, F.A.M. Current and historic gene flow of the sand goby *Pomatoschitus minutus* on the European Continental Shelf and in the Mediterranean Sea. *Biol. J. Linn. Soc.* **2004**, *83*, 561–576. [CrossRef]
47. Gysels, E.S.; Hellemans, B.; Pampoulie, C.; Volckaert, F.A.M. Phylogeography of the common goby, *Pomatoschistus microps*, with particular emphasis on the colonization of the Mediterranean and the North Sea. *Mol. Ecol.* **2004**, *13*, 403–417. [CrossRef] [PubMed]
48. Mejri, R.; Lo Brutto, S.; Hassine, O.K.B.; Arculeo, M. A study on *Pomatoschistus tortonesei* Miller 1968 (Perciformes, Gobiidae) reveals the Siculo-Tunisian Strait (STS) as a breakpoint to gene flow in the Mediterranean basin. *Mol. Phylogenet. Evol.* **2009**, *53*, 596–601. [CrossRef] [PubMed]
49. Mejri, R.; Arculeo, M.; Ben Hassine, O.K.; Lo Brutto, S. Genetic architecture of the marbled goby *Pomatoschistus marmoratus* (Perciformes, Gobiidae) in the Mediterranean Sea. *Mol. Phylogenet. Evol.* **2011**, *58*, 395–403. [CrossRef]
50. Boissin, E.; Hoareau, T.B.; Berrebi, P. Effects of current and historic habitat fragmentation on the genetic structure of the sand goby *Pomatoschistus minutus* (Osteichthys, Gobiidae). *Biol. J. Linn. Soc.* **2011**, *102*, 175–198. [CrossRef]
51. Giovannotti, M.; La Mesa, M.; Caputo, V. Life style and genetic variation in teleosts: The case of pelagic (*Aphia minuta*) and benthic (*Gobius niger*) gobies (Perciformes: Gobiidae). *Mar. Biol.* **2009**, *156*, 239–252. [CrossRef]
52. Agorreta, A.; San Mauro, D.; Schliewen, U.; Van Tassell, J.L.; Kovačić, M.; Zardoya, R.; Rüber, L. Molecular phylogenetics of Gobioidei and phylogenetic placement of European gobies. *Mol. Phylogenet. Evol.* **2013**, *69*, 619–633. [CrossRef]
53. Miller, P.J. Fishes of the North-eastern Atlantic and the Mediterranean. In *Gobiidae*; Whitehead, P.J.P., Bauchot, M.-L., Hureau, J.-C., Nielsen, J., Tortonese, E., Eds.; Unesco: Paris, France, 1986; pp. 1019–1085. ISBN 92-3-002309-4.
54. Engin, S.; Turan, D.; Kovačić, M. First record of the red-mouthed goby, *Gobius cruentatus* (Gobiidae), in the Black Sea. *Cybium* **2007**, *31*, 87–88.
55. Wilkins, H.K.A.; Myers, A.A. Shelter utilization by *Gobius cruentatus* and *Thorogobius ephippiatus* (Teleostei: Gobiidae). *J. Fish Biol.* **1993**, *43*, 763–773. [CrossRef]
56. Kovačić, M.; Golani, D. First record of three gobiid species in the Levant. *Cybium* **2007**, *31*, 89–91.
57. Kovačić, M.; Miletić, M.; Papageorgiou, N. A first checklist of gobies from Crete with ten new records. *Cybium* **2011**, *35*, 245–253.
58. Machordom, A.; Doadrio, I. Evidence of a cenozoic Betic-Kabilian connection based on freshwater fish phylogeography (*Luciobarbus*, Cyprinidae). *Mol. Phylogenet. Evol.* **2001**, *18*, 252–263. [CrossRef] [PubMed]
59. Chow, S.; Hazama, K. Universal PCR primers for S7 ribosomal protein gene introns in fish. *Mol. Ecol.* **1998**, *7*, 1255–1256. [PubMed]
60. Šanda, R.; Vukić, J.; Choleva, L.; Křížek, J.; Šedivá, A.; Shumka, S.; Wilson, I.F. Distribution of loach fishes (Cobitidae, Nemacheilidae) in Albania, with genetic analysis of populations of *Cobitis ohridana*. *Folia Zool.* **2008**, *57*, 42–50.
61. Hall, T.A. BioEdit: A user-friendly biological sequence alignment editor and analysis program for Windows 95/98/NT. *Nucleic Acids Symp. Ser.* **1999**, *41*, 95–98.
62. Darriba, D.; Taboada, G.L.; Doallo, R.; Posada, D. jModelTest 2: More models, new heuristics and high-performance computing. *Nat. Methods* **2015**, *9*, 6–9.
63. Nei, M.; Kumar, S. *Molecular Evolution and Phylogenetics*; Oxford University Press: Oxford, UK, 2000.
64. Rozas, J.; Ferrer-Mata, A.; Sánchez-DelBarrio, J.C.; Guirao-Rico, S.; Librado, P.; Ramos-Onsins, S.E.; Sánchez-Gracia, A. DnaSP 6: DNA sequence polymorphism analysis of large data sets. *Mol. Biol. Evol.* **2017**, *34*, 3299–3302. [CrossRef]
65. Excoffier, L.; Lischer, H.E.L. Arlequin suite ver 3.5: A new series of programs to perform population genetics analyses under Linux and Windows. *Mol. Ecol. Resour.* **2010**, *10*, 564–567. [CrossRef] [PubMed]
66. Tamura, K.; Stecher, G.; Peterson, D.; Filipski, A.; Kumar, S. MEGA6: Molecular evolutionary genetics analysis version 6.0. *Mol. Biol. Evol.* **2013**, *30*, 2725–2729. [CrossRef] [PubMed]

67. Stephens, M.; Smith, N.J.; Donnelly, P. A new statistical method for haplotype reconstruction from population data. *Am. J. Hum. Genet.* **2001**, *68*, 978–989. [CrossRef] [PubMed]
68. Clement, M.; Snell, Q.; Walker, P.; Posada, D.; Crandall, K. TCS: Estimating gene genealogies. In Proceedings of the 16th International Parallel and Distributed Processing Symposium (IPDPS 2002), Fort Lauderdale, FL, USA, 15–19 April 2002; Volume 2, p. 0184.
69. Leigh, J.W.; Bryant, D. POPART: Full-feature software for haplotype network construction. *Methods Ecol. Evol.* **2015**, *6*, 1110–1116. [CrossRef]
70. Mantel, N. The detection of disease clustering and a generalized regression approach. *Cancer Res.* **1967**, *27*, 209–220. [PubMed]
71. Beerli, P.; Felsenstein, J. Maximum likelihood estimation of a migration matrix and effective population sizes in n subpopulations by using a coalescent approach. *Proc. Natl. Acad. Sci. USA* **2001**, *98*, 4563–4568. [CrossRef]
72. Drummond, A.J.; Rambaut, A.; Shapiro, B.; Pybus, O.G. Bayesian coalescent inference of past population dynamics from molecular sequences. *Mol. Biol. Evol.* **2005**, *22*, 1185–1192. [CrossRef]
73. Drummond, A.J.; Suchard, M.A.; Xie, D.; Rambaut, A. Bayesian phylogenetics with BEAUti and the BEAST 1.7. *Mol. Biol. Evol.* **2012**, *29*, 1969–1973. [CrossRef]
74. Rambaut, A.; Drummond, A.J.; Xie, D.; Baele, G.; Suchard, M.A. Posterior summarisation in Bayesian phylogenetics using Tracer 1.7. *Syst. Biol.* **2018**, *67*, 901–904. [CrossRef]
75. Sefc, K.M.; Wagner, M.; Zangl, L.; Weiß, S.; Steinwender, B.; Arminger, P.; Weinmaier, T.; Balkic, N.; Kohler, T.; Inthal, S.; et al. Phylogeographic structure and population connectivity of a small benthic fish (*Tripterygion tripteronotum*) in the Adriatic Sea. *J. Biogeogr.* **2020**. [CrossRef]
76. Shanks, A.L. Pelagic larval duration and dispersal distance revisited. *Biol. Bull.* **2009**, *216*, 373–385. [CrossRef] [PubMed]
77. Robinson, R.A.; Leslie, W.G.; Theocharis, A. Mediterranean Sea Circulation. In *Encyclopedia of Ocean Sciences*, 1st ed.; Steele, J.H., Thorpe, S.A., Turekian, K.K., Eds.; Academic Press: San Diego, USA, 2001; pp. 1689–1705.
78. Serra, I.A.; Innocenti, A.M.; Di Maida, G.; Calvo, S.; Migliaccio, M.; Zambianchi, E.; Pizzigalli, C.; Arnaud-Haond, S.; Duarte, C.M.; Serrao, E.A.; et al. Genetic structure in the Mediterranean seagrass *Posidonia oceanica*: Disentangling past vicariance events from contemporary patterns of gene flow. *Mol. Ecol.* **2010**, *19*, 557–568. [CrossRef] [PubMed]
79. Thomas, L.; Jason Kennington, W.; Stat, M.; Wilkinson, S.P.; Kool, J.T.; Kendrick, G.A. Isolation by resistance across a complex coral reef seascape. *Proc. R. Soc. B Biol. Sci.* **2015**, *282*, 50–59. [CrossRef] [PubMed]
80. Munguia-Vega, A.; Marinone, S.G.; Paz-Garcia, D.A.; Giron-Nava, A.; Plomozo-Lugo, T.; Gonzalez-Cuellar, O.; Weaver, A.H.; García-Rodriguez, F.J.; Reyes-Bonilla, H. Anisotropic larval connectivity and metapopulation structure driven by directional oceanic currents in a marine fish targeted by small-scale fisheries. *Mar. Biol.* **2018**, *165*, 1–16. [CrossRef]
81. Banks, S.C.; Piggott, M.P.; Williamson, J.E.; Bové, U.; Holbrook, N.J.; Beheregaray, L.B. Oceanic variability and coastal topography shape genetic structure in a long-dispersing sea urchin. *Ecology* **2007**, *88*, 3055–3064. [CrossRef] [PubMed]
82. Benestan, L.; Quinn, B.K.; Maaroufi, H.; Laporte, M.; Clark, F.K.; Greenwood, S.J.; Rochette, R.; Bernatchez, L. Seascape genomics provides evidence for thermal adaptation and current-mediated population structure in American lobster (*Homarus americanus*). *Mol. Ecol.* **2016**, *25*, 5073–5092. [CrossRef]
83. Borges, R.; Faria, C.; Gil, F.; Gonçalves, E.J. Early Development of Gobies. In *The Biology of Gobies*; Patzner, R., Van Tassell, J.L., Kovačić, M., Kapoor, B.G., Eds.; CRC Press: Enfield, NH, USA; Taylor and Francis Group: Enfield, NH, USA; Science Publishers: Enfield, NH, USA, 2006; pp. 401–455.
84. Raventós, N.; Macpherson, E. Planktonic larval duration and settlement marks on the otoliths of Mediterranean littoral fishes. *Mar. Biol.* **2001**, *138*, 1115–1120.
85. Koblmüller, S.; Steinwender, B.; Weiß, S.; Sefc, K.M. Gene flow, population growth and a novel substitution rate estimate in a subtidal rock specialist, the black-faced blenny *Tripterygion delaisi* (Perciformes, Blennioidei, Tripterygiidae) from the Adriatic Sea. *J. Zool. Syst. Evol. Res.* **2015**, *53*, 291–299. [CrossRef]
86. Pappalardo, A.M.; Francisco, S.M.; Fruciano, C.; S Lima, C.; Pulvirenti, V.; Tigano, C.; Robalo, J.I.; Ferrito, V. Mitochondrial and nuclear intraspecific variation in the rusty blenny (*Parablennius sanguinolentus*, Blenniidae). *Hydrobiologia* **2017**, *802*, 141–154. [CrossRef]

87. Domingues, V.S.; Stefanni, S.; Brito, A.; Santos, R.S.; Almada, V.C. Phylogeography and demography of the Blenniid *Parablennius parvicornis* and its sister species *P. sanguinolentus* from the northeastern Atlantic Ocean and the western Mediterranean Sea. *Mol. Phylogenet. Evol.* **2008**, *46*, 397–402. [CrossRef] [PubMed]
88. Miller, J.A.; Shanks, A.L. Evidence for limited larval dispersal in black rockfish (*Sebastes melanops*): Implications for population structure and marine-reserve design. *Can. J. Fish. Aquat. Sci.* **2004**, *61*, 1723–1735. [CrossRef]
89. Almany, G.R.; Berumen, M.L.; Thorrold, S.R.; Planes, S.; Jones, G.P. Local replenishment of coral reef fish populations in a marine reserve. *Science* **2007**, *316*, 742–744. [CrossRef] [PubMed]
90. Kovačić, M. Unusual morphological and ecological characteristics of hyperbenthic juveniles of *Gobius cruentatus*. *J. Fish Biol.* **2004**, *65*, 545–558. [CrossRef]
91. Šegvić-Bubić, T.; Marrone, F.; Grubišić, L.; Izquierdo-Gomez, D.; Katavić, I.; Arculeo, M.; Lo Brutto, S. Two seas, two lineages: How genetic diversity is structured in Atlantic and Mediterranean greater amberjack *Seriola dumerili* Risso, 1810 (Perciformes, Carangidae). *Fish. Res.* **2016**, *179*, 271–279. [CrossRef]
92. Viñas, J.; Bremer, J.A.; Pla, C. Phylogeography of the Atlantic bonito (*Sarda sarda*) in the northern Mediterranean: The combined effects of historical vicariance, population expansion, secondary invasion, and isolation by distance. *Mol. Phylogenet. Evol.* **2004**, *33*, 32–42. [CrossRef]
93. Tougard, C.; Folly, J.; Berrebi, P. New light on the evolutionary history of the common goby (*Pomatoschistus microps*) with an emphasis on colonization processes in the Mediterranean Sea. *PLoS ONE* **2014**, *9*, e91576. [CrossRef]
94. Durand, J.D.; Blel, H.; Shen, K.N.; Koutrakis, E.T.; Guinand, B. Population genetic structure of *Mugil cephalus* in the Mediterranean and Black Seas: A single mitochondrial clade and many nuclear barriers. *Mar. Ecol. Prog. Ser.* **2013**, *474*, e91576. [CrossRef]
95. Gonzalez, E.G.; Zardoya, R. Relative role of life-history traits and historical factors in shaping genetic population structure of sardines (*Sardina pilchardus*). *BMC Evol. Biol.* **2007**, *7*, 197. [CrossRef]
96. Domingues, V.S.; Santos, R.S.; Brito, A.; Alexandrou, M.; Almada, V.C. Mitochondrial and nuclear markers reveal isolation by distance and effects of Pleistocene glaciations in the northeastern Atlantic and Mediterranean populations of the white seabream (*Diplodus sargus*, L.). *J. Exp. Mar. Bio. Ecol.* **2007**, *346*, 102–113. [CrossRef]
97. Bahri-Sfar, L.; Lemaire, C.; Ben Hassine, O.K.; Bonhomme, F. Fragmentation of sea bass populations in the western and eastern Mediterranean as revealed by microsatellite polymorphism. *Proc. R. Soc. London Ser. B, Biol. Sci.* **2000**, *267*, 929–935. [CrossRef] [PubMed]
98. Borsa, P.; Blanquer, A.; Berrebi, P. Genetic structure of the flounders *Platichthys flesus* and *P. stellatus* at different geographic scales. *Mar. Biol.* **1997**, *129*, 233–246. [CrossRef]
99. Wagner, M.; Bračun, S.; Skofitsch, G.; Kovačić, M.; Zogaris, S.; Iglésias, S.P.; Sefc, K.M.; Koblmüller, S. Diversification in gravel beaches: A radiation of interstitial clingfish (*Gouania*, Gobiesocidae) in the Mediterranean Sea. *Mol. Phylogenet. Evol.* **2019**, *139*, 106525. [CrossRef] [PubMed]
100. Rossi, A.R.; Perrone, E.; Sola, L. Genetic structure of gilthead seabream, *Sparus aurata*, in the central Mediterranean sea. *Cent. Eur. J. Biol.* **2006**, *1*, 636–647. [CrossRef]

Journal of
Marine Science and Engineering

MDPI

Article

Filling the Gap of Data-Limited Fish Species in the Eastern Mediterranean Sea: A Contribution by Citizen Science

Roxani Naasan Aga Spyridopoulou [1,*], Joachim Langeneck [2], Dimitris Bouziotis [1], Ioannis Giovos [1,2,3], Periklis Kleitou [1,3] and Stefanos Kalogirou [4,*]

1 iSea, Environmental Organisation for the Preservation of the Aquatic Ecosystems, Ochi Av. 11, 55438 Agios Paulos, Thessaloniki, Greece; dbouziotis@hotmail.com (D.B.); ioannis.giovos@isea.com.gr (I.G.); pkleitou@merresearch.com (P.K.)
2 Department of Biology, University of Pisa, via Derna 1, 56126 Pisa, Italy; jlangeneck@biologia.unipi.it
3 Marine and Environmental Research (MER) Lab Ltd., 202 Amathountos Av., Marina Gardens, Block B, Offices # 13–14, Parekklisia, 4533 Limassol, Cyprus
4 Hellenic Centre for Marine Research, Hydrobiological Station of Rhodes, 85100 Rhodes, Greece
* Correspondence: roxani.naasanaga@isea.com.gr (R.N.A.S.); stefanos.kalogirou@gmail.com (S.K.); Tel.: +30-23-1309-0696 (R.N.A.S.); +46-76-559-1708 (S.K.)

Received: 8 January 2020; Accepted: 5 February 2020; Published: 10 February 2020

check for updates

Abstract: The biodiversity of the Mediterranean Sea is rapidly changing due to anthropogenic activity and the recent increase of seawater temperature. Citizen science is escalating as an important contributor in the inventory of rare and data-limited species. In this study, we present several records of five data-limited native fish species from the eastern Mediterranean Sea: *Alectis alexandrina* (Geoffroy Saint-Hilaire, 1817), *Ranzania laevis* (Pennant, 1776), *Dalatias licha* (Bonnaterre, 1788), *Lophotus lacepede* (Giorna, 1809), and *Sudis hyalina* (Rafinesque, 1810). All of the records were collected by a participatory process involving fishers and validated by associated taxonomic experts of the citizen science programme "Is it Alien to you? Share it!!!". This study fills an important gap for the distribution of the reported species and signifies the important role of citizen participation as a tool for extending marine biodiversity knowledge and fisheries management in an area with several gaps of knowledge on targeted and non-targeted species.

Keywords: Alexandria pompano; Slender sunfish; Kitefin shark; Crested oarfish; Barracudina; eastern Mediterranean Sea

1. Introduction

The Mediterranean Sea is facing several unprecedented anthropic pressures (e.g. pollution, habitat destruction, and geographical reshuffling of species) [1,2]. Along with climate change, species community shifts are regularly observed, leading to the tropicalization of the Mediterranean Sea [3,4]. Projections have indicated that at least 25% of the Mediterranean continental shelf might experience a total modification of species assemblages by the end of the 21st century [5]. Major gaps exist regarding deep species assemblages [6], particularly rare and data-limited species. This is mainly attributed to fragmented research, scarcity of observations, less fishing pressure, and practical difficulties in monitoring deeper waters.

Historically, the low economic value of non-targeted species has led to less data and a lack of vulnerability assessments. Some progress has been made in recent years to develop methods on status and risk assessments [7], such as mixed fisheries [8] and policy requirements [9,10]. Thus, the methodology has been developed, including time-series catch data [11], life history aspects [12], and size structure [13]. Another aspect is the willingness by citizens to pay for climate adaptation and fisheries resources, as shown by Tulone et al. [14].

Citizen science is emerging as a key component for the exploration of marine biodiversity, being widely acknowledged by scientists, policy-makers, and conservationists, due to its capacity to address conservation issues that are related to rare species, climate change, and coastal systems [15–17]. Currently, there is an increasing number of citizen science efforts in the Mediterranean Sea t focused on various topics that are related to the marine environment [18–20]. This provides high potential for addressing data gaps related to least studied species. However, citizen science data should be carefully treated, since they often incorporate taxonomic uncertainties and misidentification due to photographic identification [21,22].

Here, we report several records of five data-limited fish species in the eastern Mediterranean Sea and highlight the use of citizen science as an emerging scientific tool for increasing our understanding on species distributions and contributing to fisheries management in the Mediterranean Sea. In addition, we used current published literature to highlight and give a quantitative aspect of current and historical knowledge of the studied species distribution in the eastern Mediterranean Sea (Supplementary Table S1).

2. Materials and Methods

The citizen-science programme "Is it Alien to you? Share it!!!" was initially launched in May 2016 by the Environmental Organisation iSea, with the aim of recording information on the occurrence, distribution, and expansion of marine non-native and rare species in Greece. An online easy-to-upload data repository was generated for citizens that allowed for records with pictures to be tracked. In addition, a Facebook group was created and all of the uploaded pictures to the group were automatically uploaded to the same data repository. At the time, this manuscript was written, this project's Facebook group numbers > 10,000 members, among which approximately 5000 are actively engaged. Each observer is requested to provide for each photo information on species size in total length (TL in cm.) and/or wet weight (WW; in g.). It is also requested to provide more information regarding the depth (D; in meters), the number of individuals (N), the location in latitude and longitude (LAT: LON), the date (DD: MM: YYYY), and the type of observation, as: underwater observation (UW), stranded (S) (i.e. found deceased on shore). Trained observers with silhouettes measured all of the individuals that were categorized as UW. If fishing gear/practice was provided, each observation was categorized following the FAO classification: longline (LL), bottom longline (bLL), surface longline (sLL), harpoons (HAR), and handlines and pole-lines that are hand-operated (LHP) (Figure 1, Table 1). All of the records are pre-scanned for quality and sent out for external evaluation to taxonomic experts prior to the inclusion in the final dataset of iSea. Only confirmed observations are recorded in the dataset, either through photo-identification or in-situ validation.

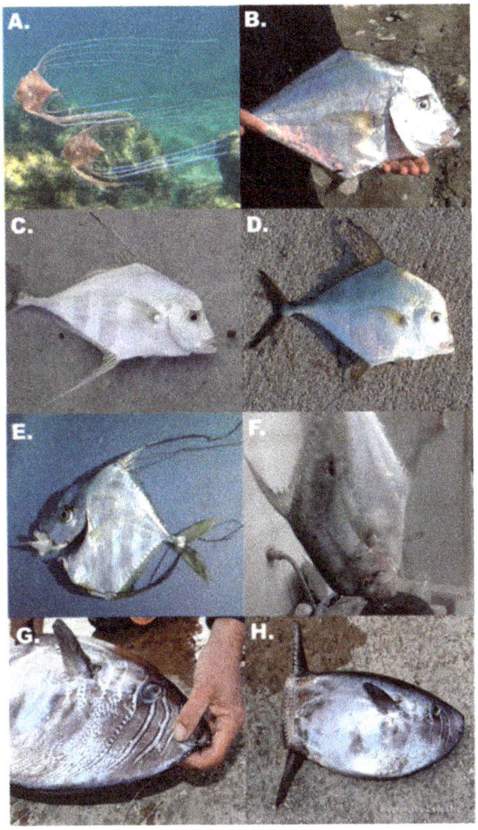

Figure 1. (**A**,**B**): *Alectis alexandrina* juvenile sighted in Protaras, Cyprus; (**B**,**C**): *Alectis alexandrina* adults caught in Zygi and Limassol port (Cyprus) respectively; (**D**): *Alectis alexandrina* adult captured in the harbour of Kos Island, Greece; (**E**): *Alectis alexandrina* juvenile sighted in Zygi, Cyprus; (**F**): *Alectis alexandrina* captured by spearfishing in Akrotiri Bay, Cyprus; (**G**,**H**): *Ranzania laevis* individual captured with surface longline in Psathoura Island, Greece.

Table 1. Records of five rare species in the eastern Mediterranean Sea with information on number of individuals (N), depth (m.), total length (cm.), area, coordinates (lat: lon) in decimal degrees and type of observation (underwater, UW; shore fishing, SF; spearfishing,; demersal longline, LL, Otter bottom trawl OTB).

n	Species	N Individuals	Depth	TL	Area	Coordinates	Type of Observation
1	*A. alexandrina*	2	N/A	10	Protaras, Cyprus	35°00′48.6″ N, 34°02′14.6″ E	UW
2	*A. alexandrina*	1	N/A	45	Zygi, Cyprus	34°43′35.3″ N, 33°20′15.0″ E	SF
3	*A. alexandrina*	1	5	45	Old port of Limassol, Cyprus	34°40′13.8″ N, 33°02′34.8″ E	SF
4	*A. alexandrina*	1	N/A	50	Port of Kos Island, Greece	36°53′43.8″ N, 27°17′19.0″ E	SF
5	*A. alexandrina*	1	N/A	22	Zygi, Cyprus	34°43′41.5″ N, 33°20′25.8″ E	S
6	*A. alexandrina*	1	5	60	Akrotiri Bay, Cyprus	34°33′20.2″ N, 33°00′59.8″ E	SF
7	*A. alexandrina*	1	4	15	Ayia Napa, Cyprus	34°58′53.0″ N, 34°00′17.9″ E	SF

Table 1. *Cont.*

n	Species	N Individuals	Depth	TL	Area	Coordinates	Type of Observation
8	*A. alexandrina*	1	N/A	45	Zygi, Cyprus	34°40′15.7″ N, 33°02′39.4″ E	S
9	*A. alexandrina*	1	3,5	6,5	Protaras, Cyprus	35°00′30.1″ N, 34°03′48.4″ E	UW
10	*R. laevis*	1	200	53	Psathoura island, Greece	39°26′46.0″ N, 24°04′46.1″E	LL
11	*D. licha*	1	612	150	Levitha island, Greece	36°58′23.2″ N, 26°29′11.4″ E	LL
12	*D. licha*	1	580	150	Amorgos island, Greece	36°46′25.7″ N, 26°02′38.0″ E	LL
13	*L. lacepede*	1	500	160	Between Skopelos and Euboea island, Greece	36°46′25.7″ N, 23°34′44.0″ E	LL
14	*L. lacepede*	1	500	140	Between Skopelos and Euboea island, Greece	36°46′25.7″ N, 23°34′44.0″ E	LL
15	*L. lacepede*	1	540	60	Between Santorini and Anafi island, Greece	36°22′39.4″ N, 25°35′44.5″ E	LL
16	*L. lacepede*	1	300	100	Chalkidiki Peninsula, Greece	39°44′10.7″ N, 23°22′56.6″ E	LL
17	*L. lacepede*	1	540	140	Crete island, Greece	35°34′58.0″ N, 23°29′29.1″ E	LL
18	*L. lacepede*	1	40	25	Rhodes island, Greece	36°27′21.6″ N, 28°13′05.5″ E	BS
19	*S. hyalina*	1	700	N/A	Kastellorizo island, Greece	35°51′41.4″ N, 30°06′32.0″ E	LL
20	*S. hyalina*	1	200	30	Akrotiri Bay, Cyprus	34°32′46.0″ N, 32°56′56.4″ E	SF
21	*S. hyalina*	1	100	N/A	Crete island, Greece	35°31′20.8″ N, 24°00′55.5″ E	BS
22	*S. hyalina*	1	300	40	Morfou Gulf, Cyprus	35°13′43.4″ N, 32°52′01.1″ E	BS
23	*S. hyalina*	1	400	N/A	Crete island, Greece	34°54′55.1″ N, 24°54′50.6″ E	BS
24	*S. hyalina*	1	100	N/A	Saronikos Gulf, Athens	37°50′12.0″ N, 23°18′50.7″ E	BS
25	*S. hyalina*	1	612	35	Levitha island, Greece	37°00′02.9″ N, 26°27′32.6″ E	LL

All of the data used in this study has been uploaded to an electronic repository (https://wp.me/P94Vaj-1Mm).

3. Results

3.1. *Alectis Alexandrina*

On 20th November 2012, two juvenile individuals (≈ 10 cm TL) were sighted and photographed near Protaras, Cyprus (Levant Sea, 35°00′48.6″ N, 34°02′14.6″ E) (Figure 1A; Table 1). On 4th February 2015, an individual with a TL of 45 cm was caught by a recreational fisherman at Zygi, Cyprus (Levant Sea, 34°43′35.3″ N, 33°20′15.0″ E), while a second individual, approximately of the same size, was caught on 30th May 2015 in the old port of Limassol (Levant Sea, 34°40′13.8″ N, 33°02′34.8″ E) (Figure 1B,C). On July 2016, a single individual with a TL of ≈ 50 cm and 1100 g in weight was caught by a recreational fisherman at the port of Kos Island (Aegean Sea; 36°53′43.8″ N, 27°17′19.0″ E; Figure 1D). On 16th March 2017, a juvenile individual (≈ 22 cm TL for 100 g) was found stranded near to Zygi village, Cyprus (Levant Sea, 34°43′41.5″ N, 33°20′25.8″ E) (Figure 1E). On 11th October 2017, an individual (≈ 60 cm TL) was captured by spearfishing at a depth of 5 m at Akrotiri Bay in Limassol (Levant Sea, 34°33′20.2″ N, 33°00′59.8″ E). During 2018 there had been two more records that were collected on 9th of October an individual of ≈ 15 cm TL by a spear fisher and another individual ≈ 6.5 cm TL was photographed by a diver; both occurrences were in Cyprus Ayia Napa, (Levant Sea, 34°58′53.0″ N, 34°00′17.9″ E) and Portaras Bay, (Levant Sea, 35°00′30.1″ N, 34°03′48.4″ E), accordingly (Table 1). Finally, there has been another stranded individual of 45 cm TL, in Zygi, Cyprus (Levant Sea, 34°40′15.7″ N, 33°02′39.4″ E), (Table 1; Figure 4).

3.2. *Ranzania Laevis*

On 21st December 2014, an individual of *R. laevis* was captured with a surface longline near Psathoura Island in North Aegean Sea (39°26′46.0″ N, 24°04′46.1″ E) at a depth of approximately 200 m and landed at Skopelos Island (Figure 1G, H). The specimen was 53 cm in TL. The fisherman and a local journalist contacted the local authorities, which identified the species as "Propela", the common Greek name for *Mola mola*. On 29th of July 2017 the journalist contacted iSea, in the context of the citizen science project "Is it Alien to you? Share it!!!" for sharing a log of rare species observations from Skopelos Island. Among the observations, the record of *R. laevis* was found and all of the available information was collected.

3.3. *Dalatias Licha*

On 5th November 2016, a single male individual of *D. licha* (≈ 150 cm TL) was caught by a demersal longline at a depth of 580 m off Amorgos Island (Aegean Sea, 36°46′25.7″ N, 26°02′38.0″ E) (Figure 2A). On 29th August 2019 an individual was caught off Levithas island at a depth of 612 m in depth by a demersal longline over muddy substrate (Aegean Sea; 36°58′23.2″ N, 26°29′11.4″ E) (Table 1; Figure 4).

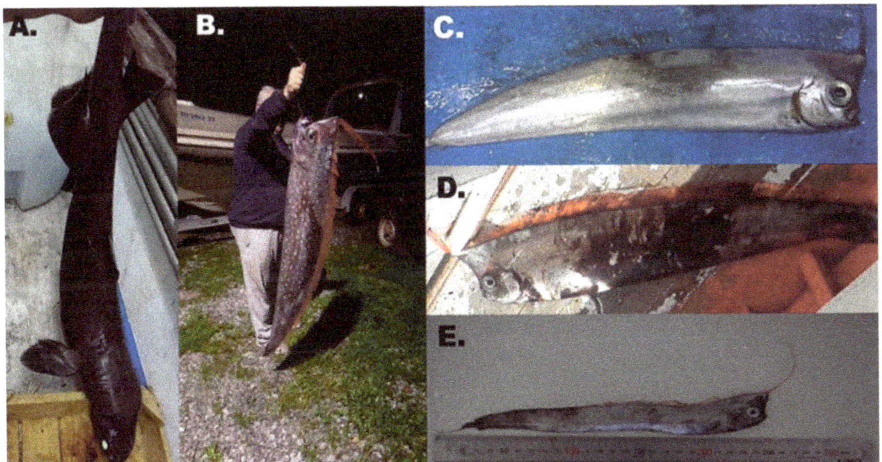

Figure 2. (**A**): Male *Dalatias licha* caught with demersal longlines off Amorgos Island, Greece; (**B**). *L. lacepede* caught with demersal longlines between Skopelos and Euboea Islands, Greece; (**C**): *L. lacepede* caught with demersal longlines between Santorini and Anafi Islands, Greece; and, (**D**): *L. lacepede* caught off the Coast of Chalkidiki Peninsula, Greece; (**E**): *L. lacepede*, Rhodes Island, Greece.

3.4. *Lophotus Lacepede*

Demersal longlines set at 500 m deep bottoms between Skopelos and Euboea Islands (Aegean Sea, 36.7738° N, 23.5789° E) allowed for the capture of a large individual (≈ 160 cm TL, 10 kg WW) (Figure 2B) on 3rd March 2017. A few days later, on 9th March 2017, a slightly smaller individual (approximately 140 cm in TL and 8 kg in WW) was collected from the same area while using the same fishing gear. A third, distinctly smaller individual (≈ 60 cm in TL) (Figure 2C) was captured by demersal longlines at a depth of 540 m between Santorini and Anafi Islands (Aegean Sea; 36.3776° N, 25.5957° E). Lastly, on 6th October 2017, a fourth large individual (approximately 100 cm in TL and 6kg in WW) (Figure 2D) was captured by a longline targeting swordfish off the coast of Chalkidiki Peninsula towards Sporades Islands (North Aegean Sea; 39°44′10.7″N, 23°22′56.6″ E) at a depth of approximately 300 m. On September 12, 2017 an individual of ≈ 25 cm in TL was caught outside Rhodes Island, Greece (36°27′21.6″ N, 28°13′05.5″ E) (Figure 2E). On 1st of November 2018, an individual of ≈ 140 cm in TL was caught by a professional fisher with a demersal longline at a depth of 540 m (Cretan Sea, 23°29′29.1″ N, 35°34′58.0″ E) (Figure 4).

3.5. *Sudis Hyalina*

On 1st July of 2014, an individual was caught at 100 m in depth by demersal longline of a recreational fisher, off Crete Island (Aegean Sea, 35°31′20.8″N, 24°00′55.5″ E). On 6th November 2016 an individual was caught by demersal longline approximately 30 miles south-east from Kastellorizo Island and at approximately 700 m in depth (Aegean Sea; 35°51′41.4″N, 30°06′32.0″E) (Figure 3A). The specimen was found mutilated by another fish, possibly a scabbardfish (*Lepidopus caudatus;* Euphrasen, 1788), but features of the head allowed for a uniequivocal identification. On 17th August 2017, another individual of *S. hyalina* (≈ 30 cm TL) was captured at Akrotiri, Cyprus (34°32′46.0″N, 32°56′56.4″ E) by a recreational fisher at an approximate depth of 200 m (Figure 3B,C). During 2018, another record was added from Morfou in Cyprus; the individual was approximately 40 cm in TL and it was caught at 300 m in depth (Levantine Sea, 35°13′43.4″ N, 32°52′01.1″ E). On 19th May 2019, another individual was caught in 400 m by shore-based fishing off Crete island (Cretan Sea, 34°54′55.1″ N, 24°54′50.6″ E); on 29th August 2019, an individual was caught off Levithas island at 612 m in depth with demersal longline over muddy substrate along with a *D. licha* individual (Aegean Sea, 37°00′02.9″ N, 26°27′32.6″ E), (Table 1; Figure 4). The latest record was on 17th September 2019, where another individual of ≈ 20 cm in TL was caught by demersal longline at 100 m in depth at Saronikos Gulf (Aegean Sea, 37°50′12.0″ N, 23°18′50.7″ E).

Figure 3. (**A**): Head of *Sudis hyalina* collected from demersal longlines off Kastellorizo Island, Greece; (**B**) individual captured in Akrotiri Bay, Cyprus; and, (**C**): individual captured in Chalkidiki Peninsula, Greece.

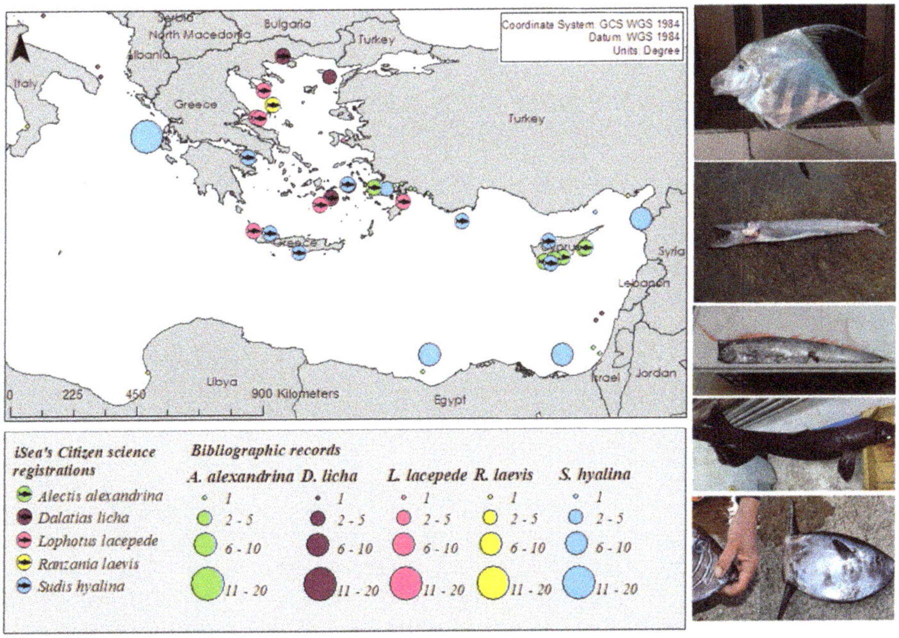

Figure 4. Geographic density dependent bubble plot on five uncommon data-limited species (*Alectis alexandrina, Dalatias licha, Lophotus lacepede, Ranzania laevis* and *Sudis hyalina*) in the eastern Mediterranean Sea.

We found that, based on current literature (Supplementary Table S1), the reported number of individuals in this study contributed 21.6% of the reports the species in the eastern Mediterranean Sea and 27.3% to the total number of reports during the last decade. For some species, the contribution to the total number of reports in the eastern Mediterranean was higher: (a) *A. alexandrina* contributed 34.4% of the reports the species in the eastern Mediterranean Sea and 35.7% to total number of reported species in the eastern Mediterranean Sea during the last decade, (b) *R. laevis* contributed 11.1% of the reports of the species in the eastern Mediterranean Sea and 20% of the reports of the species in the eastern Mediterranean Sea during the last decade, (c) *D. licha* contributed 18.9% of the reports of the species in the eastern Mediterranean Sea and 28.6% of the reports of the species in the eastern Mediterranean Sea during the last decade, (d) *L. lacepede* contributed 46.1% of the reports of the species in the eastern Mediterranean Sea and 75% of the reports of the species in the eastern Mediterranean Sea during the last decade and (e) *S. hyalina* contributed 11.9% of the reports of the species in the eastern Mediterranean Sea and 14% of the reports of the species in the eastern Mediterranean Sea during the last decade.

4. Discussion

The present work provides additional information on five uncommon fish species for the Aegean and Levantine Sea. Rare species are often considered to be data-limited. However, spatial and temporal variations in the distribution of rare species might provide relevant signs of climate and environmental change [23,24]. Until recently, the scarce communication between researchers and citizens entailed the loss of an important part of available information on these species, while the recent increase in the use of social networks allowed for a closer communication that can result in an increase of records of rare and non-native species [20,25–28]. To date, the project "Is it Alien to you? Share it!!!" has gathered a vast amount of information regarding the distribution and establishment of species, including several first records, as well as expansion evidences [20].

Alectis alexandrina is a thermophilic species with a rather sporadic occurrence in the Mediterranean Sea. Fishers state that the species is increasingly caught the last years, being very common in Cyprus and rather sporadic in southern Aegean Sea, despite the few observations from the northeast Mediterranean Sea.

Ranzania laevis commonly known as slender sunfish (Fam. Molidae) is a pelagic-oceanic cosmopolitan species found in tropical and temperate seas and feeding on small fish, planktonic crustaceans, and jellyfish. It is one of the two Molidae species occurring in the Mediterranean Sea; however, reports of *R. laevis* from the eastern basin are scarce. Occasional records from Greece [29], Cyprus and Israel [30], Turkey [31], and Lybia [32] have been reported. Current distribution of *R. laevis* covers the whole Mediterranean Sea while quantitative information has revealed a rather sporadic occurrence and local peaks of abundance with increased zooplankton biomass [23]. The reported species is the second record from the north Aegean Sea [29].

The remaining three species reported are typical rare deep-water species in the eastern Mediterranean Sea [6]. Recent studies demonstrated the occurrence of these deep-water species, as previously overlooked in the eastern Mediterranean Sea [33,34].

Dalatias licha, commonly known as the kitefin shark, is one of the largest deep-sea sharks occurring in the Mediterranean Sea, being distributed at depths between 200 and 900 m, where it can be considered a top predator. It is although an uncommon species in the western and central part of the basin [35,36], while its presence in the eastern Mediterranean Sea is considered as very scarce [37,38], signifying any reports as highly valuable, also given the conservation status of the Mediterranean population of the species as vulnerable [39].

Lophotus lacepede, commonly known as the crested oarfish, is a large bathypelagic species that is remarkably sporadic in the whole Mediterranean Sea. Similarly, to the majority of deep-sea fish species it is more regularly observed in the western Mediterranean Sea, with only a few records in the eastern part of the basin. Currently, only four published records are known, all being from the Aegean Sea [40]. The observations that are reported in the current work highlight the deep waters between Chalkidiki Peninsula and Evoia Islands as an important area for the species.

Sudis hyalina, which is a rare bathypelagic cosmopolitan fish species, is considered to be regular in the western part of the Mediterranean basin, while only five documented records of the species exist from the eastern Mediterranean basin [41]. In the Aegean Sea only two records of the species have been published [29]. Consequently, the record presented in this study is the third record of the species from the Aegean Sea. Interestingly, the report from Cyprus is the first record of the species from the Cypriot territorial waters, most possibly an overlooked species.

5. Conclusions

This study reveals that the contribution of citizen-science accounts for 45.4% of the total number of the studied species published records and 58.5% of the total number of published records during the last decade in the eastern Mediterranean Sea. This signifies the important role of citizen-science efforts, both for scientific and public awareness. Nine more records of the reported species have been published since "Is it Alien to you? Share it!!!" was launched [38,42–46], whereas, in this article, we report the occurrence of 26 individuals contributing with 74.2% during the same period. Of course, it can be hard to determine whether the species density in the region has increased or if the reports were overlooked before pictures of the species were circulated on social media as something rare. Thus, post-hoc interviews with citizens and fishermen who report catches can be used to reconstruct species distributions and densities.

Supplementary Materials: The following are available online at http://www.mdpi.com/2077-1312/8/2/107/s1, Table S1. Up-to-date records of five data-limited species in the eastern Mediterranean Sea.

Author Contributions: Conceptualization, S.K., I.G., P.K. and R.N.A.S.; methodology, S.K., R.N.A.S.; software, R.N.A.S.; validation, S.K. and R.N.A.S.; formal analysis, S.K. and R.N.A.S.; investigation, S.K.; R.N.A.S., I.G., P.K., J.L., D.B.; resources, S.K.; R.N.A.S., I.G., P.K., J.L., D.B.; data curation, R.N.A.S., S.K.; writing—original draft preparation, S.K.; R.N.A.S., I.G., P.K., J.L., D.B.; writing—review and editing, S.K., R.N.A.S., I.G., P.K., J.L. and D.B.; visualization, S.K., R.N.A.S.; supervision, S.K.; project administration, I.G.; funding acquisition, I.G. All authors have read and agreed to the published version of the manuscript.

Funding: This research was funded by iSea, Environmental Organisation for the preservation of the aquatic ecosystems—"Is it Alien to you? Share it!!!".

Acknowledgments: We are grateful to the fisherman Nikolas Katsillis (Greece) for sharing his pictures and data for three specimens of *S. hyalina, L. lacepede* and *D. licha*, Antonis Voutsas (Greece) for sharing his picture and data of *L. lacepede*, to Werner Wolf (Austria), Pabos Charalambous (Cyprus), Savvas Ioannou (Cyprus), Michalis Ilia (Cyprus) and Antonis Salachoris (Greece) for sharing data and pictures of the *A. alexandrina* specimens, Angelos Kouriefs (Cyprus) for sharing data and pictures of *S. hyalina* specimen and Georgios Poulios (Greece) for sharing data and pictures of *R. laevis* and all the other citizen scientists who contributed in gathering the information needed.

Conflicts of Interest: The authors declare no conflict of interest. The funders had no role in the design of the study; in the collection, analyses, or interpretation of data; in the writing of the manuscript, or in the decision to publish the results.

References

1. Fortibuoni, T.; Idighieri, F.; Giovanardi, O.; Pranovi, F.; Zucchetta, M. Climate impact on Italian fisheries (Mediterranean Sea). *Reg. Environ. Chang.* **2015**, *15*, 931–937. [CrossRef]
2. Macías, D.; Castilla-Espino, D.; García-del-Hoyo, J.J.; Navarro, G.; Catalán, I.A.; Renault, L.; Ruiz, J. Consequences of a future climatic scenario for the anchovy fishery in the Alboran Sea (SW Mediterranean): A modeling study. *J. Mar. Syst.* **2014**, *135*, 150–159. [CrossRef]
3. Occhipinti-Ambrogi, A. Global change and marine communities: Alien species and climate change. *Mar. Pollut. Bull.* **2007**, *55*, 342–352. [CrossRef] [PubMed]
4. Lejeusne, C.; Chevaldonné, P.; Pergent-Martini, C.; Boudouresque, C.F.; Perez, T. Climate change effects on a miniature ocean: The highly diverse, highly impacted Mediterranean Sea. *Trends Ecol. Evol.* **2010**, *25*, 250–260. [CrossRef] [PubMed]
5. Lasram, B.R.F.; Guilhaumon, F.; Albouy, C.; Somot, S.; Thuiller, W.; Mouillot, D. The Mediterranean Sea as a 'cul-de-sac' for endemic fishes facing climate change. *Glob. Chang. Biol.* **2010**, *16*, 3233–3245. [CrossRef]
6. Danovaro, R.; Company, J.B.; Corinaldesi, C.; D'Onghia, G.; Galil, B.; Gambi, C.; Gooday, A.J.; Lampadariou, N.; Luna, G.M.; Morigi, C.; et al. Deep-sea biodiversity in the Mediterranean Sea: The known, the unknown, and the unknowable. *PLoS ONE* **2010**, *5*, e11832. [CrossRef]
7. Thorson, J.T.; Kell, L.T.; De Oliveira, J.A.A.; Sampson, D.B.; Punt, A.E. Development, testing, and evaluation of data-poor assessment and fisheries management methods. *Fish. Res.* **2015**, *171*, 1–200. [CrossRef]
8. Ford, R.B.; Galland, A.; Clark, M.R.; Crozier, P.; Duffy, C.A.J.; Dunn, M.R.; Francis, M.P.; Wells, R. *Qualitative (Level 1) Risk Assessment of the Impact of Commercial Fishing on New Zealand Chondrichthyans*; New Zealand Aquatic Environment and Biodiversity Report; Ministry for Primary Industries: Wellington, New Zealand, 2015; Volume 157, 111p.
9. Hobday, A.J.; Smith, A.D.M.; Stobutzki, I.C.; Bulman, C.; Daley, R.; Dambacher, J.M.; Deng, R.A.; Dowdneya, J.; Fullera, M.; Furlania, D.; et al. Ecological risk assessment for the effects of fishing. *Fish. Res.* **2011**, *108*, 372–384. [CrossRef]
10. Berkson, J.; Barbieri, L.; Cadrin, S.; Cass-Calay, S.; Crone, P.; Dorn, P.; Friess, C.; Kobayashi, D.; Miller, T.J.; Patrick, W.S.; et al. *Calculating Acceptable Biological Catch for Stocks That Have Reliable Catch Data Only (Only Reliable Catch Stocks—ORCS)*; NMFS-SEFSC-616; NOAA Technical Memorandum: Miami, FL, USA, 2011; 56p.
11. Zhou, S.; Punt, A.E.; Ye, Y.; Ellis, N.; Dichmont, C.M.; Haddon, M.; Smith, D.C.; Smith, A.D.M. Estimating stock depletion level from patterns of catch history. *Fish Fish.* **2017**, *18*, 742–751. [CrossRef]
12. Dick, E.J.; MacCall, A.D. Depletion-based stock reduction analysis: A catch-based method for determining sustainable yields for data-poor fish stocks. *Fish. Res.* **2011**, *110*, 331–341. [CrossRef]
13. Hordyk, A.; Ono, K.; Sainsbury, K.; Loneragan, N.; Prince, J. Some explorations of the life history ratios to describe length composition, spawning-per-recruit, and the spawning potential ratio. *ICES J. Mar. Sci.* **2015**, *72*, 204–216. [CrossRef]

14. Tulone, A.; Crescimanno, M.; Vrontis, D.; Galati, A. Are coastal communities able to pay for the protection of fish resources impacted by climate change? *Fish. Res.* **2020**, *221*, 105374. [CrossRef]
15. Thiel, M.; Penna-Díaz, M.A.; Luna-Jorquera, G.; Salas, S.; Sellanes, J.; Stotz, W. Citizen scientists and marine research: Volunteer participants, their contributions, and projection for the future. *Oceanogr. Mar. Biol. Annu. Rev.* **2014**, *52*, 257–314. [CrossRef]
16. Theobald, E.J.; Ettinger, A.K.; Burgess, H.K.; DeBey, L.B.; Schmidt, N.R.; Froehlich, H.E.; Parrish, J.K. Global change and local solutions: Tapping the unrealized potential of citizen science for biodiversity research. *Biol. Conserv.* **2015**, *181*, 236–244. [CrossRef]
17. Cousins, J.A.; Huxham, M.; Winton, D. Using citizen science to address conservation issues related to climate change and coastal systems. In *Science for Coastal and Marine Conservation*; Cigliano, J.A., Ballard, J.A., Eds.; Routledge: London, UK, 2017; p. 20.
18. Gerovasileiou, V.; Dailianis, T.; Panteri, E.; Michalakis, N.; Gatti, G.; Sini, M.; Dimitriadis, C.; Issaris, Y.; Salomidi, M.; Filiopoulou, I.; et al. CIGESMED for divers: Establishing a citizen science initiative for the mapping and monitoring of coralligenous assemblages in the Mediterranean Sea. *Biodiver. Data J.* **2016**, *4*, e8692. [CrossRef]
19. Barash, A.; Salingre, S.; Grosmark, Y.; Rothman, S.; Stoilas, V.S.; Maximiadi, M.; Tuncer, S.; Lapinski, M.; Nuez, I.; Bakiu, R.; et al. The MECO project (Mediterranean Elasmobranch Citizen Observations): Creating a large-scale database of elasmobranchs observations using social media. In Proceedings of the 22nd Annual European Elasmobranch Association Meeting, Peniche, Portugal, 12–14 October 2018.
20. Giovos, I.; Kleitou, P.; Poursanidis, D.; Batjakas, I.; Bernardi, G.; Crocetta, F.; Doumpas, N.; Kalogirou, S.; Kampouris, T.E.; Keramidas, I.; et al. Citizen-science for monitoring marine invasions and stimulating public engagement: A case project from the eastern Mediterranean. *Biol. Invasions* **2019**, *21*, 3707–3721. [CrossRef]
21. Marshall, A.D.; Pierce, S.J. The use and abuse of photographic identification in sharks and rays. *J. Fish. Biol.* **2012**, *80*, 1361–1379. [CrossRef]
22. Katsanevakis, S.; Moustakas, A. Uncertainty in Marine Invasion Science. *Front. Mar. Sci.* **2018**, *5*, 38. [CrossRef]
23. Dulčić, J.; Beg Paklar, G.; Grbec, B.; Morović, M.; Matić, E.; Lipej, L. On the occurrence of the ocean sunfish *Mola mola* and slender sunfish *Ranzania laevis* in the Adriatic Sea. *J. Mar. Biol. Assoc. UK* **2007**, *87*, 789–796. [CrossRef]
24. Caballero-Alfonso, A.; Ganzedo-López, U.; Díez-Díez, G.; Castro, J. New record of Ocythoe tuberculata (Cephalopoda: Ocythoidae) in the north-east Atlantic related to sea warming. *Mar. Biodivers. Rec.* **2009**, *2*, e79. [CrossRef]
25. Andaloro, F.; Castriota, L.; Falautano, M.; Azzurro, E.; Deidun, A.; Fenech-Farrugia, A. Public feedback on early warning initiatives undertaken for hazardous non-indigenous species: The case of *Lagocephalus sceleratus* from Italian and Maltese waters. *Manag. Biol. Invasion* **2016**, *7*, 313–319. [CrossRef]
26. Kletou, D.; Hall-Spencer, J.M.; Kleitou, P. A lionfish (*Pterois miles*) invasion has begun in the Mediterranean Sea. *Mar. Biodivers. Rec.* **2016**, *9*, e46. [CrossRef]
27. Bariche, M.; Azzurro, E. *Enhancing Early Detection through Social Networks: A Facebook Experiment*; Rapports et Proces-Verbaux Reunions Commission International Mer Mediterranee; Reunions Commission International: Monaco, 2016; Volume 41, p. 413.
28. Langeneck, J.; Marcelli, M.; Bariche, M.; Azzurro, E. Social networks allow early detection of non indigenous species: First record of the red drum *Sciaenops ocellatus* (Actinopterygii: Perciformes: Sciaenidae) in Italian waters. *Acta Adriat.* **2017**, *58*, 365–370. [CrossRef]
29. Papaconstantinou, C. *Fauna Graeciae. An Updated Checklist of the Fishes in the Hellenic Seas*; Monographs on Marine Sciences, 7; HCMR: Athens, Greece, 2014; p. 340.
30. Tortonese, E. *Molidae.* In *Fishes of the North-Eastern Atlantic and the Mediterranean*; Whitehead, P.J.P., Bauchot, M.-L., Hureau, J.-C., Nielsen, J., Tortonese, E., Eds.; UNESCO: Paris, France, 1986; pp. 1348–1350.
31. Bilecenoglu, M.; Taskavak, E.; Mater, S.; Kaya, M. Checklist of the marine fishes of Turkey. *Zootaxa* **2002**, *113*, 1–194. [CrossRef]
32. Elbaraasi, H.; Elsalini, O. Occurrence of slender sunfish *Ranzania laevis* Pennant, 1776, (*Molidae*) in the eastern part of the Libyan coast (southern Mediterranean). *Turk J. Fish Aquat. Sci.* **2010**, *10*, 549–550. [CrossRef]
33. Goren, M.; Galil, B.S. A checklist of the deep sea fishes of the Levant Sea, Mediterranean Sea. *Zootaxa* **2015**, *3994*, 507–530. [CrossRef]

34. Farrag, M.M.S. Deep-sea ichthyofauna from Eastern Mediterranean Sea, Egypt: Update and new records. *Egypt. J. Aquat. Res.* **2016**, *42*, 479–489. [CrossRef]
35. Ragonese, S.; Vitale, S.; Dimech, M.; Mazzola, S. Abundances of demersal sharks and chimaera from 1994-2009 scientific surveys in the central Mediterranean Sea. *PLoS ONE* **2013**, 8. [CrossRef]
36. Navarro, J.; López, L.; Coll, M.; Barría, C.; Saez-Liante, R. Short- and long-term importance of small sharks in the diet of the rare deep-sea shark *Dalatias licha*. *Mar. Biol.* **2014**, *161*, 1697–1707. [CrossRef]
37. Corsini-Foka, M. Uncommon fished from Rhodes and nearby marine region (SE Aegean Sea, Greece). *J. Biol. Res. Thessaloniki* **2009**, *12*, 125–133.
38. Ergüden, D.; Çekic, M.; Alagoz Ergüden, S.; Altun, A.; Uygur, N. Occurrence of adult female Kitefin shark *Dalatias licha* (Bonnaterre, 1788) in Iskenderun Bay (Eastern Mediterranean, Turkey). *Commagene J. Biol.* **2017**, *1*, 60–62. [CrossRef]
39. Walls, R.H.L.; Guallart, J. *Dalatias licha*; The IUCN Red List of Threatened Species 2016: E.T6229A16527825. IUCN, 2016. Available online: http://www.iucnredlist.org/details/6229/16527825 (accessed on 10 February 2020).
40. Minos, G.; Karidas, T.; Economidis, P.S. Range extension for crested oarfish *Lophotus lacepede* Giorna, 1809 in the waters of the northern Aegean Sea, Greece. *Acta Adriat.* **2015**, *56*, 283–290.
41. Türker, D.; Kara, A.; Bal, H.; Tünay, Ö.K. Occurrence of rare deep-water fish *Sudis hyalina* Rafinesque, 1810 (Paralepididae) in Gökova Bay, Aegean Sea of Turkey. *J. Appl. Ichthyol.* **2017**, *33*, 535–538. [CrossRef]
42. Ceyhan, T.; Akyol, O. Additional record of *Alectis alexandrina* (Carangidae) from the north-western Levantine Sea (Fethiye, Turkey). *Ann. Ser. Hist. Nat.* **2016**, *26*, 221–223. [CrossRef]
43. Goren, M.; Galil, B.S.; Diamant, A.; Stern, N.; Levitt-Barmats, Y. Invading up the food web? Invasive fish in the southeastern Mediterranean Sea. *Mar. Biol.* **2016**, *163*, 180. [CrossRef]
44. Chatzispyrou, A.; Aroni, M.; Lefkaditou, E.; Kapiris, K.; Giovos, I.; Anastasopoulou, A. Some Biological Information on a Female Kitefin Shark, *Dalatias licha* (Bonnaterre, 1788) Stranded in the Laconikos Gulf of Greece (SE Ionian Sea). *Turk. J. Fish. Aquat. Sci.* **2019**, *19*, 1069–1072. [CrossRef]
45. Yapici, S. New and additional records of rare fish species from the Anatolian coasts of Turkey. *Mugla J. Sci. Technol.* **2019**, *5*, 13–15. [CrossRef]
46. Ergüden, D.; Gürlek, M.; Ergüden, S.A.; Ayas, D.; Altun, A. Presence of *Ranzania laevis* (Pennant, 1776) in Iskenderun Bay, Turkey. *Thalassas Int. J. Mar. Sci.* **2019**. [CrossRef]

Journal of
Marine Science and Engineering

MDPI

Article

Biologic Aspects of Juveniles of the Common Stingray, *Dasyatis pastinaca* (Linnaeus, 1758) (Elasmobranchii, Dasyatidae), from the Central Mediterranean Sea

Francesco Tiralongo [1,2,3,*], Giuseppina Messina [1] and Bianca Maria Lombardo [1]

1 Department of Biological, Geological and Environmental Sciences, University of Catania, 95124 Catania, Italy; giuseppina.messina@unict.it (G.M.); bm.lombardo@unict.it (B.M.L.)
2 Ente Fauna Marina Mediterranea, 96012 Avola, Italy
3 GRIS, Gruppo Ricercatori Italiani sugli Squali, Razze e Chimere, 16132 Genova, Italy
* Correspondence: francesco.tiralongo@unict.it

Received: 24 February 2020; Accepted: 7 April 2020; Published: 10 April 2020

check for updates

Abstract: Data on the biology of *Dasyatis pastinaca* are absent from the Ionian Sea and only a few studies were conducted in the Mediterranean Sea. Some biological and ecological aspects of *D. pastinaca* were investigated between November 2019 and February 2020 in the central Mediterranean Sea. In particular, we investigated several morphologic, population and ecological aspects of the species. The analysis of the stomach contents showed that *D. pastinaca* is a generalist carnivorous, mainly feeding on small crustaceans and polychaetes. The Levin's index value (B_i) was 0.85. The sex ratio showed no significant differences from 1:1 ratio. Females were larger than males, but no statistical differences were found in disc width-weight and total length-disc width relationships between sexes. Most of the specimens caught were juveniles and inhabit shallow sandy bottoms.

Keywords: eastern Sicily; Batoidea; elasmobranchs; diet; coastal fishery

1. Introduction

Elasmobranchs are key top predators in most marine environments and play an essential role in regulating and structuring marine ecosystems [1,2]. On the other hand, due to their low fecundity and delayed age at maturity, elasmobranchs are highly vulnerable to fishing activity and are often affected by high by-catch rates [3,4]. All this is reflected in the dramatic decline of shark and ray populations and, as demonstrated by several researches conducted in the Mediterranean Sea over the last decades, most species of elasmobranchs have dramatically declined in number mainly due to overfishing and illegal fishing [5–9].

In Italian seas, stingrays (Dasyatidae) are represented by 4 species and 3 genera [10]: *Dasyatis* Rafinesque, 1810, *Pteroplatytrygon* Fowler, 1910 and *Taeniura* Müller & Henle, 1837. The genus *Dasyatis* comprises *Dasyatis pastinaca* (Linnaeus, 1758) and *Dasyatis centroura* (Mitchill, 1815). These two species are very similar but can be distinguished by the presence (*D. pastinaca*) or absence (*D. centroura*) of a dorsal keel behind the spine on the tail and by the presence of quite developed spines and tubercles on the dorsal surface of the body and tail in large specimens of *D. centroura* [11]. Furthermore, between all the Mediterranean species of the genus *Dasyatis*, *D. pastinaca* is the most abundant and widely distributed one [12]. *Pteroplatytrygon violacea* (Bonaparte, 1832) is a less common species than *D. pastinaca*, while *Taeniura grabata* (Geoffroy Saint-Hilaire, 1817) is very rare and was only occasionally caught in the Mediterranean, and only once in Italian waters [10,13]. Because of their low or no commercial value, stingrays are usually discarded [14–16]. However, due to their vulnerability to fishing activities, the conservation and correct management of these vulnerable K-selected species require special efforts to ensure their perpetuation.

The common stingray, *D. pastinaca*, is an Atlantic-Mediterranean species whose distribution extends from the southwestern Baltic Sea and throughout the Mediterranean Sea to Senegal [17]. It is listed as "data deficient" in the IUCN Red List [18]. The common stingray inhabits soft bottoms of coastal waters, down to a depth of about 200 m [19,20]. It is commonly caught with trammel nets and trawls and usually discarded [14–16]. The maximum recorded disc width (DW) is 64 cm, but common sizes range from 20 to 40 cm DW. The survival rate of this species, when caught with trammel net, is equal or close to 0 [16]. Indeed, because of their defensive behavior, fishermen are forced to kill the stingrays with the aim of avoiding serious injuries from their venomous spine on the tail. Another noteworthy fact of this species is the antimicrobial and anti-proliferative effects of its skin mucus [21].

Data on the biology and ecology of *D. pastinaca* in the Mediterranean Sea are scarce, and no data are available from the Ionian Sea. Therefore, the aim of this study is to provide new data on the biology and ecology of the common stingray, providing the first data from specimens from the Ionian Sea. In particular, we investigated size frequency distribution, sex ratio, disc width-weight relationships, total length-disc width relationships, diet composition and habitat of *D. pastinaca* from specimens caught as discards in trammel nets targeting cuttlefish, *Sepia officinalis* Linnaeus, 1758, in the central Mediterranean Sea.

2. Materials and Methods

A total of 120 specimens of *D. pastinaca* were collected along the coastline extending for about 21 km in the southeast coast of Sicily (Ionian Sea), from Avola to Marzamemi (Figure 1). The specimens were caught by professional fishermen at 5–30 m depth between the 10 November 2019 and 10 February 2020 with trammel nets targeting cuttlefish (*S. officinalis*) [16]. Trammel nets were deployed overnight (from 6 pm to 4 am) for about 10 h, on sandy and mixed bottoms (sand and rocks), close to *Posidonia oceanica* seagrass meadows. In order to better represent the population, specimens were randomly selected from different fishing vessels operating in the area. Data about bottom nature and depth of capture were collected through interviews to fishermen.

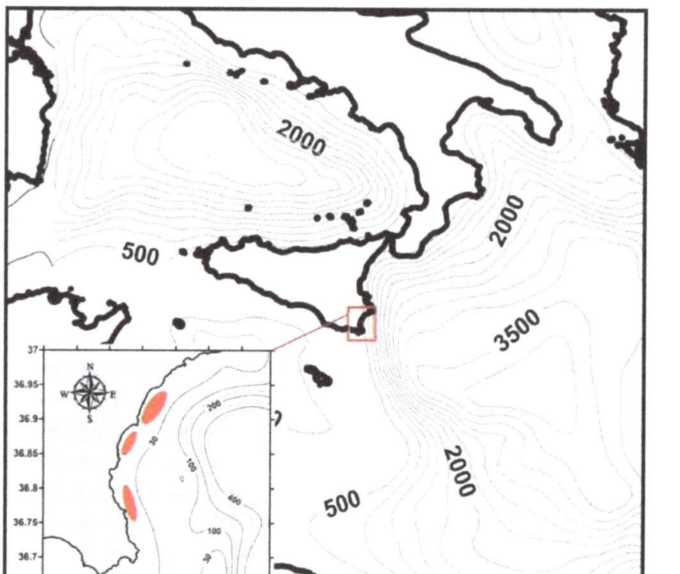

Figure 1. Study area (indicated in red) in the Ionian coast of Sicily (central Mediterranean Sea).

Each specimen was weighed and measured (disc width and total length). The sex was determined by the presence (male) or absence (female) of claspers. Weight and disc width measures were used for the disc width-weight relationships following the formula: $W = aDW^b$, where W is the weight in grams (g), DW is the disc width in centimeters (cm), a is the intercept and b is the slope of the regression curve. Total length (TL) and disc width measures were used for the total length-disc width relationships following the formula: $DW = yTL + x$, where y is the slope and x is the intercept of the regression line. Disc-width frequency distributions were constructed for both sexes. A chi-square test was used to verify if there was a significant difference ($\alpha = 0.05$) between the observed and the expected sex ratio (M:F, 1:1) of the whole sample. To test if the regressions of the weight (W) on disc width (DW) were significantly different ($\alpha = 0.05$) for the two sexes, and to test if the regressions of the disc width on total length (TL) were significantly different ($\alpha = 0.05$) for the two sexes, an analysis of covariance (ANCOVA) was employed. Five outliers (females > 2143 g) were excluded from the analyses.

The stomach was removed from each fish as soon as possible after landing, and its content analyzed. All the prey items in the stomachs were counted, washed in clean seawater and dried with blotter paper, identified under a microscope to the lowest taxonomic level possible and weighed to the nearest 0.01 g.

The frequency of occurrence (%F), percentage weight (%W), percentage abundance (%N) and the Index of Relative Importance (%IRI) were calculated for each prey category [22,23]. The vacuity index (percentage of empty stomachs) was also calculated.

According to the value of their percentage abundance (%N), prey were grouped into three categories [24]: dominant (N > 50%), secondary (10% < N < 50%) and accidental (N < 10%).

The feeding strategy of *D. pastinaca* was visually examined using a modified version of the Costello (1990) graph [25] by plotting the prey-specific biomass (P*i*) against their frequency of occurrence (%F) [26]:

$$Pi = \frac{SWi}{SWt_i} \times 100 \tag{1}$$

where P*i* is the prey-specific biomass of prey *i*, SW_i the stomach content biomass of prey *i*, and SWt_i the total stomach content biomass in those predators with prey *i* in the stomach.

Standardized Levin's index (B*i*) was used to evaluate the breadth of the diet [27]:

$$B = \frac{1}{\Sigma p_j^2} \tag{2}$$

$$Bi = \frac{B-1}{B_{max}-1} \tag{3}$$

where p_j is the relative frequency specimens in the jth prey item and B_{max} is the total number of prey item categories found. B*i* is comprised between 0 and 1. The more the value of B*i* is close to 0, the narrower is the trophic niche of the species investigated. Conversely, the closer the value of B*i* is to 1, the wider the trophic niche. Hence, if $Bi < 0.40$, the species is considered a "specialist feeder"; if $0.40 < Bi < 0.60$, the species is considered an "intermediate feeder"; if $Bi > 0.60$ the species is considered a "generalist feeder" [28].

A cumulative prey curve [29] was computed with R Studio [30] in order to evaluate whether the number of analyzed stomachs was sufficient to describe the diet of the species. The estimated number of prey groups with the associated SD were plotted against the cumulative number of individuals whose stomach was examined.

3. Results

Out of the total 120 specimens sampled, 57 (47.5%) were males and 63 (52.5%) were females. The sex ratio did not deviate significantly from 1:1 (M:F, 0.9:1; *p*-value > 0.05). Females were on average larger than males, ranging from 31 to 76.4 cm in total length, from 16.8 to 45.5 cm in disc width and from 144 to 3781 g in weight (Figure 2, Table 1). However, the ANCOVA test did not indicate significant differences (p-value > 0.05) in disc width-weight relationships between sexes and neither did total length-disc width relationships (p-value > 0.05). The disc width-weight relationships for both sexes and combined showed a positive allometry (Table 2, Figure 3). The parameters of the linear regression of the total length-disc width relationships were: DW = 0.641TL−4.2508 for males; DW = 0.641TL−4.122 for females; and DW = 0.641TL−4.1867 for combined sexes (Figure 4).

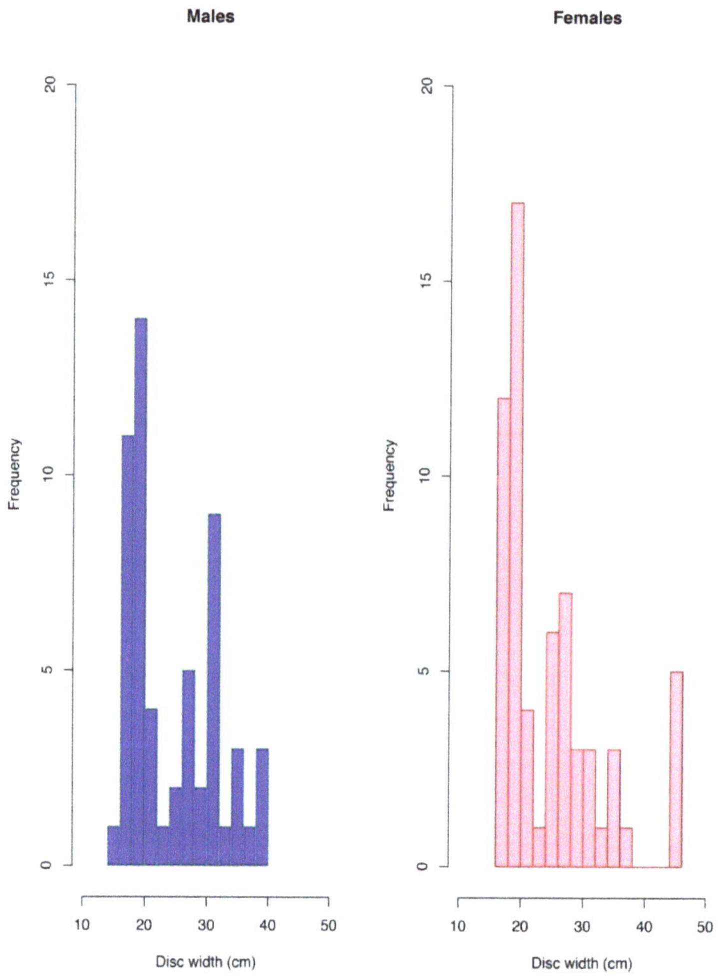

Figure 2. Sex composition by size classes of *Dasyatis pastinaca*.

Table 1. Sex distribution, total length, disc width and weight of *Dasyatis pastinaca* in the Ionian Sea (central Mediterranean Sea). The sex ratio (M:F, 0.9:1) was not significantly different from 1:1 (chi-square: $p > 0.05$).

SEX	N	Total Length (TL) cm			Disc Width (DW) cm			Weight (W) g		
		Range	Mean	SD	Range	Mean	SD	Range	Mean	SD
M	57	31.2–62.8	44.62	10.72	15.9–38.2	24.36	6.94	125–2195	672.1	616.51
F	63	31–76.4	44.99	12.55	16.8–45.5	24.72	8.09	144–3781	750.1	952.76
Combined	120	31–76.4	44.82	11.67	15.9–45.5	24.55	7.53	125–3781	713	808.29

Table 2. Disc width-weight relationships parameters of *D. pastinaca* in the Ionian Sea (central Mediterranean Sea); C.I. = 95% confidence interval.

SEX	N	a	C.I. a	b	C.I. b	R^2	*p*-Value
M	57	0.0133	0.0095–0.0185	3.303	3.197–3.409	0.99	<0.01
F	63	0.0163	0.0126–0.0217	3.229	3.147–3.311	0.99	<0.01
Combined	120	0.0149	0.0121–0.0183	3.262	3.196–3.327	0.99	<0.01

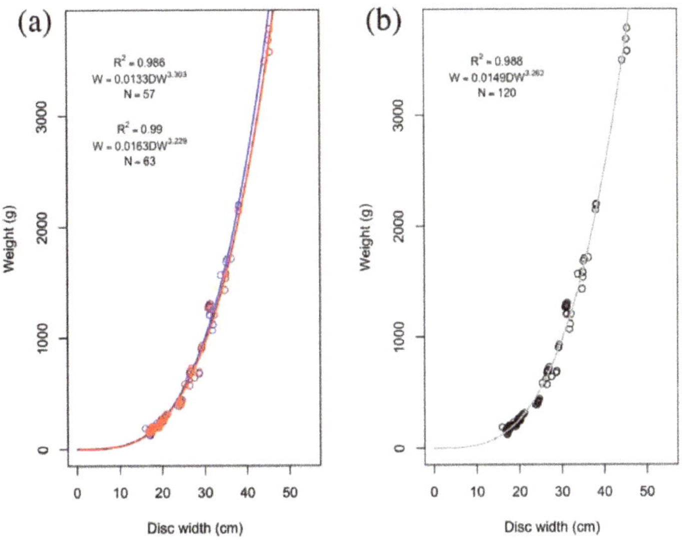

Figure 3. Disc width-weight relationships of *D. pastinaca* for both sexes (**a**) and combined (**b**). A: in red for females (N = 63) and blue for males (N = 57).

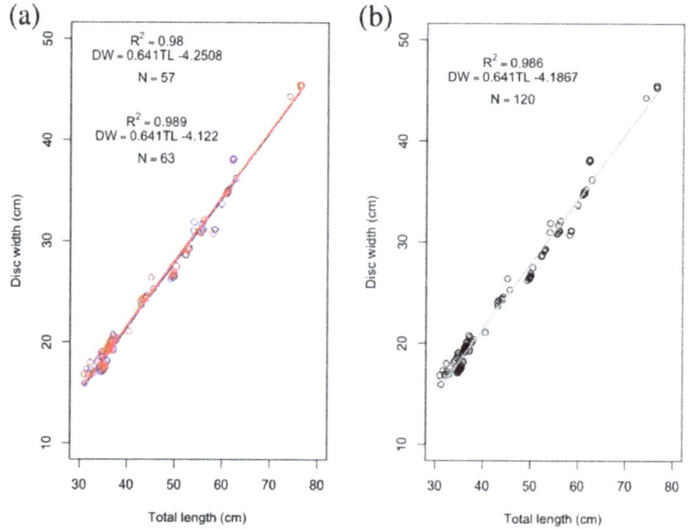

Figure 4. Disc width-length relationships of *D. pastinaca* for both sexes (**a**) and combined (**b**). A: in red for females (N = 63) and blue for males (N = 57).

Considering the size at first maturity (L_{50} = 62.5 cm and 43 cm in DW for males and females, respectively) reported by a study conducted in the nearby Aegean Sea [31], a high percentage of specimens (virtually all males and the majority of females) were juveniles. All the specimens were caught on sandy bottom and within the whole bathymetric range investigated (5–30 m). A total of only 5 specimens (all females > 2143 g) were of relatively large size.

The cumulative prey curve showed that relatively few stomachs (about 50) are necessary for a reliable description of the diet of *D. pastinaca* (Figure 5). Out of the 120 specimens analyzed, only 67 had prey in their stomach (vacuity index of 55.83%). The stomach content analysis revealed that juveniles of *D. pastinaca* mainly feed on small benthic crustaceans and also, to a lesser extent, polychaetes (Table 3). Fish represented a negligible part of the diet. However, no preference for a particular prey was observed: the value of the standardized Levin's index (B*i*) was 0.85 (the digested category was included in the analysis), indicating a wide trophic niche. Most of the prey types were included into the category of "accidental prey" (N < 10%) (Table 3) and only a few (Amphipoda, Mysida and Lumbrineridae) were classified as "secondary prey" (10% < N < 50%). No dominant (N > 50%) prey types were found between the 13 identified items. Most of the prey were represented by small benthic invertebrates inhabiting sandy bottoms (Table 3). The values obtained for %IRI, %F and %W also indicated no clear dominance of any prey (Table 3). The same was clearly indicated by the plotted results of the main prey-specific biomass (P*i*) against the frequency of occurrence (%F), being in agreement with results of the other previous analysis (Figure 6).

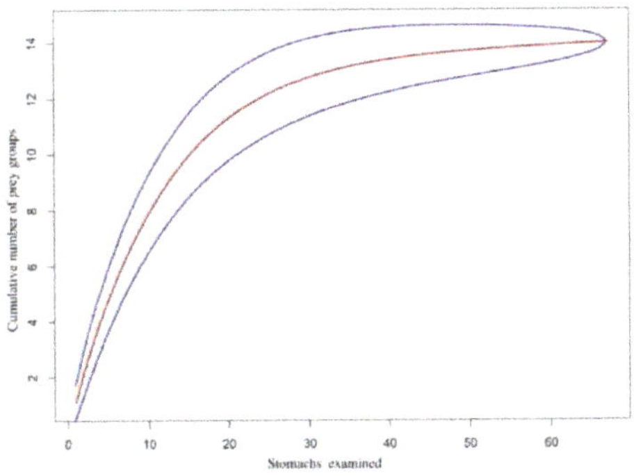

Figure 5. Cumulative prey curve (in red) as a function of sample size for all stomachs analyzed of *D. pastinaca*. Standard deviation (SD) represented with blue lines.

Table 3. Diet composition of *D. pastinaca* from the Ionian Sea (central Mediterranean Sea). %F = percentage frequency of occurrence; %N = percentage in number; %W = percentage in biomass; IRI = Index of Relative Importance of prey items and its percentage (%IRI). Values > 10% are in bold.

Food Items	%F	%N	%W	IRI	%IRI
CRUSTACEA					
Alpheus sp.	5.97	2.06	1.37	20.47	1.1
Amphipoda	7.46	**11.86**	2.07	103.91	5.6
Caridea	**14.93**	6.7	**12.32**	283.96	**15.2**
Crangon crangon	8.96	3.61	3.15	60.56	3.2
Liocarcinus sp.	**10.45**	3.61	**20.2**	248.81	**13.3**
Mysida	**10.45**	**43.81**	3.04	489.58	**26.2**
Portunus hastatus	8.96	3.09	**17.44**	183.94	9.8
Xantho sp.	4.48	1.55	8.74	46.09	2.5
MOLLUSCA					
Donacidae	7.46	3.09	2.32	40.35	2.2
Gastropoda	5.97	2.06	6.79	52.83	2.8
POLYCHAETA					
Lumbrineridae	8.96	**11.34**	**15.5**	240.48	**12.9**
Polychaeta	**10.45**	5.15	2.27	77.53	4.1
TELEOSTEI					
Pomatoschistus sp.	1.49	0.52	3.21	5.55	0.3
Digested	4.48	1.55	1.58	14.02	0.8

Figure 6. Prey-specific biomass (P*i*) plotted against the frequency of occurrence (%F) of the main prey items for *D. pastinaca* from the Ionian Sea (central Mediterranean Sea).

4. Discussion

With the exception of a few large female specimens, all or most of the individuals were juveniles. The sex ratio (statistically not significantly different from 1:1) of the fish caught during the study period indicates that there is no sexual segregation in juveniles of *D. pastinaca*.

The disc width-weight relationships showed a positive allometric growth in both sexes. Population structure and growth parameters, such as length-weight relationships are valuable measurements in order to evaluate the health conditions of fish populations [32] and are useful for comparisons of life histories of species among regions [33]. Also, total length-disc width relationships are valuable measurements to investigate intra and interspecific variation among populations [34]. Females showed larger sizes than males, but no significant differences were found in disc width-weight and total length-disc width relationships between the two sexes. Males and females attained smaller sizes than those of specimens from the Aegean Sea and Cilician Basin [31,35]. On the contrary, they reached similar size to those from the Iskenderun Bay [36]. Size differences appears to be related to the sampling depth. Indeed, while our specimens and those from the Iskenderun Bay were caught in shallow waters (5–30 and 0–50 m, respectively), the specimens from the other two Mediterranean areas (Aegean Sea and Cilician Basin) were collected in deeper waters (>50 m, down to a depth of 500 m). In consideration of this, we can speculate that juveniles of *D. pastinaca* use the study area, consisting of sandy bottoms of shallow waters close to *Posidonia oceanica* meadows, as a feeding ground. A similar result was obtained for juveniles of the rough ray, *Raja radula* Delaroche, 1809 [37].

On the basis of the number of specimens collected during the study period—and especially from interviews with local fishermen—*D. pastinaca* can be considered common in shallow waters of the Ionian coast of Sicily (central Mediterranean Sea). We also point out that in the investigated area *D. pastinaca* represented about 19% of the total elasmobranch catches in trammel nets targeting *S. officinalis* [16]. In the southeastern Sicily, the species is sympatric with other common species of elasmobranchs: *R. radula* and *Torpedo torpedo* (Linnaeus, 1758). However, while fishermen caught both adults and juveniles of *T. torpedo* [38], in *R. radula*, as in the case of *D. pastinaca*, most of the specimens were juveniles [37]. Furthermore, considering the different feeding habits between *D. pastinaca* and *T. torpedo*, their trophic niche does not overlap, resulting in little or no direct competition between these two species. Indeed, while *D. pastinaca* mainly feeds on small benthic crustaceans, as also demonstrated in this study, *T. torpedo* is a generalist piscivore [38,39]. The different diets of the two species (*D. pastinaca* and *T. torpedo*) can be explained by a different predation strategy. Indeed, only *T. torpedo* is capable of producing electric discharges to stun prey before swallowing them, and therefore this ambush predator feeds on more mobile and larger prey such as fish. Conversely, between *D. pastinaca* and *R. radula* the diet overlap was consistent. Both species mainly feed on small benthic crustaceans, with *R. radula* being a more selective feeder [37]. Hence, the differences in the diets of these species and, consequently, the different trophic competition between them, could be the cause, at least in part, of the different abundances of the species in the area: ~19%, ~33% and ~47% of the total catch of elasmobranchs, respectively for *D. pastinaca*, *R. radula* and *T. torpedo* [16].

Crustaceans (total percentage in number > 75%), followed by polychaetes, were the dominant prey. Considering the IRI, the main prey among crustaceans were represented by mysid shrimps, *Liocarcinus* sp. and Caridea. Other studies from other Mediterranean areas also showed similar results. In particular, Yeldan et al. (2009) [35] showed that the diet of specimens from Cilician coasts (Turkey) was mainly composed by crustaceans. Another study from the Black Sea (Turkey) [40] showed a similar result and similar results were also reported from the northeastern Atlantic (Azores) and from the eastern Adriatic Sea (Croatia) [41,42]. Furthermore, although the crustacean species composition varies between our and the above mentioned studies, some decapods species seem to be quite common in the diet of *D. pastinaca* and were reported from two or more of the above mentioned studies (including our study): *Alpheus glaber* (Olivi, 1792), *Crangon crangon* (Linnaeus, 1758), *Liocarcinus depurator* (Linnaeus, 1758) and *Upogebia pusilla* (Petagna, 1792). In all cases, the high value of the Levin's index (0.85) and the plot of the prey-specific biomass (*Pi*) indicated that *D. pastinaca* is a generalist feeder.

In the marine environment, sharks and batoids often occupy the highest trophic levels of the food chain [43]. The key ecological role of these species is however threatened by human activities such as illegal fishing, overfishing and habitat destruction [1], which have in some cases led to dramatic shift in relative abundance, sometimes resulting in a great increase in smaller, lower-trophic level species. However, despite the importance of understanding the feeding relationships and the energy transfer in marine ecosystems, little is known about the feeding ecology of most elasmobranch species, and this is particularly true for batoids, which have received considerably less attention than sharks [44,45]. The importance of stingray predation in regulating the abundance and composition of prey in shallow waters has been demonstrated by several authors [46,47]. Hence, there is a need to protect these vulnerable and important key species. It therefore appears of great importance in management and conservation strategies such as, for example, the establishment of marine reserve, to have a good understanding of the trophic differences and relationships among species. Hence, we also need to deepen the knowledge about the trophic ecology of these species [48].

In conclusion, our study provides new data on the biology and ecology of *D. pastinaca* in the central Mediterranean Sea (where the only available data were from the eastern Adriatic Sea), and first data from the Ionian Sea. *Dasyatis pastinaca* juveniles were found common in sandy bottoms of the investigated area. Juveniles are quite active feeders, and mainly feed on small benthic invertebrates, with a general preference for decapod crustaceans and polychaetes. However, there was no clear preference of any particular species. Additional targeted studies are needed to provide new useful data for the management and protection of this potentially vulnerable k-selected species. Still little is known about several biologic and ecological aspects of *D. pastinaca*, such as population structure, abundance, distribution and habitat selection and no data are available from the western part of the Mediterranean Sea. The experience and knowledge of fishery workers is a valid method for data collection [49,50]; a close collaboration between scientists and fishermen can lead to a better understating of the ecology and distribution of elasmobranchs.

Finally, particular attention is needed concerning the impacts of coastal fisheries on this and other coastal elasmobranch species. Considering the relatively high percentage of juveniles of the species of batoids caught in the area, a reduction of fishing pressure on these resources should be encouraged. A successful strategy could be the release of juvenile and adult specimens that are still alive and in good conditions in nearby selected areas immediately after landing.

Author Contributions: Conceptualization, F.T.; methodology, F.T. and G.M.; software, F.T.; validation, B.M.L.; formal analysis, F.T. and G.M.; investigation, F.T. and G.M.; data curation, F.T., G.M. and B.M.L.; writing—original draft preparation, F.T.; writing—review and editing, F.T., G.M and B.M.L.; supervision, B..M.L. All authors have read and agreed to the published version of the manuscript.

Funding: This research received no external funding.

Acknowledgments: We are grateful to the fishermen of Avola and Marzamemi for their support and willingness to freely share the information they had and for their help with the collection of the specimens.

Conflicts of Interest: The authors declare no conflict of interest.

References

1. Stevens, J.D.; Bonfil, R.; Dulvy, N.K.; Walker, P.A. The effects of fishing on sharks, rays and chimaeras (chondrichthyans), and the implications for marine ecosystems. *ICES J. Mar. Sci.* **2000**, *57*, 476–494. [CrossRef]
2. Bornatowski, H.; Navia, A.F.; Braga, R.R.; Abilhoa, V.; Corrêa, M.F.M. Ecological importance of sharks and rays in a structural foodweb analysis in southern Brazil. *ICES J. Mar. Sci.* **2014**, *71*, 1586–1592. [CrossRef]
3. Ragonese, S.; Cigala Fulgosi, F.; Bianchini, M.L.; Norrito, G.; Sinacori, G. Annotated check list of the skates (Chondrichthyes, Rajidae) in the Strait of Sicily (central Mediterranean Sea). *Biol. Mar. Mediterr.* **2003**, *10*, 874–881.
4. White, W.T.; Blaber, S.J.M.; Craig, J.F. The current status of elasmobranchs: Biology, fisheries and conservation. *J. Fish Biol.* **2012**, *80*, 897–900. [CrossRef] [PubMed]
5. Soldo, A. Status of sharks in the Mediterranean. *Ann. Ser. Hist. Nat.* **2003**, *13*, 191–200.

6. Cavanagh, R.D.; Gibson, C. *Overview of the Conservation Status of Cartilaginous Fishes (Chondrichthyans) in the Mediterranean Sea*; IUCN: Gland, Switzerland; Malaga, Spain, 2007; Volume VI, p. 42.
7. Bradai, M.N.; Saidi, B.; Enajjar, S. *Elasmobranchs of the Mediterranean and Black Sea: Status, Ecology and Biology Bibliographic Analysis*; FAO: Rome, Italy, 2012; Volume X, p. 103.
8. Giovos, I.; Stoilas, V.O.; Al-Mabruk, S.A.A.; Doumpas, N.; Marakis, P.; Maximiadi, M.; Moutopoulos, D.; Kleitou, P.; Keramidas, I.; Tiralongo, F.; et al. Integrating local ecological knowledge, citizen science and long-term historical data for endangered species conservation: Additional records of angel sharks (Chondrichthyes: Squatinidae) in the Mediterranean Sea. *Aquat. Conserv.* **2019**, *29*, 881–890. [CrossRef]
9. Giovos, I.; Arculeo, M.; Doumpas, N.; Katsada, D.; Maximiadi, M.; Mitsou, E.; Paravas, V.; Aga-Spyridopoulou, R.N.; Stoilas, V.O.; Tiralongo, F.; et al. Assessing multiple sources of data to detect illegal fishing, trade and mislabeling of elasmobranchs in Greek markets. *Mar. Policy* **2020**, *112*. [CrossRef]
10. Vacchi, M.; Serena, F. Chondrichthyes. *Biol. Mar. Mediterr.* **2010**, *17*, 642–648.
11. Notabartolo di Sciara, G.; Bianchi, I. *Guida Degli Squali e Delle Razze del Mediterraneo*; Franco Muzzio: Padova, Italy, 1998; p. 388.
12. Özbek, E.Ö.; Çardak, M.; Kebapçioglu, T. Spatio-temporal patterns of abundance, biomass and length-weight relationships of *Dasyatis* species (Pisces: Dasyatidae) in the Gulf of Antalya, Turkey (Levantine Sea). *J. Black Sea Medit. Environ.* **2015**, *21*, 169–190.
13. Serena, F.; Silvestri, R.; Voliani, A. Su una cattura accidentale di *Taeniura grabata* (E. Geoffroy Saint-Hilaire, 1817) (Chondrichthyes, Dasyatidae). *Biol. Mar. Mediterr.* **1999**, *6*, 617–618.
14. Relini, G.; Biagi, F.; Serena, A.; Belluscio, M.T.; Spedicato, P.; Rinelli, P.; Follesa, M.C.; Piccinetti, C.; Ungaro, N.; Sion, L.; et al. I Selaci pescati con lo strascico nei mari italiani. *Biol. Mar. Mediterr.* **2000**, *7*, 347–384.
15. Scacco, U.; Andaloro, F.; Campagnuolo, S.; Castriota, L.; Vacchi, M. Cartilaginous fishes as a component of trawl discard in Strait of Sicily (Elasmobranch Fisheries—Oral). *NAFO SCR Doc* **2002**, *2*, 87.
16. Tiralongo, F.; Messina, G.; Lombardo, B.M. Discards of elasmobranchs in a trammel net fishery targeting cuttlefish, *Sepia officinalis* Linnaeus, 1758, along the coast of Sicily (central Mediterranean Sea). *Reg. Stud. Mar. Sci.* **2018**, *20*, 60–63. [CrossRef]
17. Froese, R.; Pauly, D. FishBase. Available online: http://www.fishbase.org (accessed on 15 January 2020).
18. Serena, F.; Notabartolo di Sciara, G.; Ungaro, N. The IUCN Red List of Threatened Species. 2009. Available online: https://www.iucnredlist.org/species/161453/5427586 (accessed on 12 February 2020).
19. Whitehead, P.J.P.; Bauchot, M.L.; Hureau, J.C.; Nielsen, J.; Tortonese, E. *Fishes of the North-Eastern Atlantic and Mediterranean*; Unesco: Paris, France, 1984; Volume 1-III, p. 1473.
20. Fischer, W.; Bauchot, M.L.; Schneider, M. *Fiches FAO D'indetification des Espèces Pour le Besoins de la Peche, Méditerranée et Mer Noire*; FAO: Rome, Italy, 1987.
21. Fuochi, V.; Li Volti, G.; Camiolo, G.; Tiralongo, F.; Giallongo, C.; Distefano, A.; Petronio, G.; Barbagallo, I.; Viola, M.; Furneri, P.M.; et al. Antimicrobial and anti-proliferative effects of skin mucus derived from *Dasyatis pastinaca* (Linnaeus, 1758). *Mar. Drugs* **2017**, *15*, 342. [CrossRef] [PubMed]
22. Hyslop, E.J. Stomach content analysis, a review of methods and their application. *J. Fish. Biol.* **1980**, *17*, 411–429. [CrossRef]
23. Carrassòn, M.; Matallanas, J.; Casadevall, M. Feeding strategies of deep-water morids on the western Mediterranean slope. *Deep Sea Res.* **1997**, *44*, 1685–1699. [CrossRef]
24. N'Da, K. Regime alimentaire du rouget de roche *Mullus surmuletus* (Mullidae) dans le nord du golfe de Gascogne. *Cybium* **1992**, *16*, 159–168.
25. Costello, M.J. Predator feeding strategy and prey importance: A new geographical analysis. *J. Fish. Biol.* **1990**, *36*, 261–263. [CrossRef]
26. Amundsen, P.A.; Gabler, H.M.; Staldvik, F.J. A new approach to graphical analysis of feeding strategy from stomach contents data-modification of the Costello (1990) method. *J. Fish. Biol.* **1996**, *48*, 607–614. [CrossRef]
27. Krebs, J.C. *Ecological Methodology*; Harper & Row: New York, NY, USA, 1989; p. 620.
28. Novakowski, G.C.; Hahn, N.S.; Fugi, R. Diet seasonality and food overlap of the fish assemblage in a pantanal pond. *Neotrop. Ichthyol.* **2008**, *6*, 567–576. [CrossRef]
29. Brown, S.C.; Bizzarro, J.J.; Cailliet, G.M.; Ebert, D.A. Breaking with tradition: Redefining measures for diet description with a case study of the Aleutian skate *Bathyraja aleutica* (Gilbert 1869). *Environ. Biol. Fish.* **2012**, *95*, 3–20. [CrossRef]

30. R Development Core Team. R: A Language and Environment for Statistical Computing. Available online: http://www.R-project.org (accessed on 30 November 2018).
31. Yigin, C.C.; Ismen, A. Age, growth and reproduction of the common stingray, *Dasyatis pastinaca* from the North Aegean Sea. *Mar. Biol. Res.* **2012**, *8*, 644–653. [CrossRef]
32. Jones, R.E.; Petrell, R.J.; Pauly, D. Using modified length-weight relationships to assess the condition of fish. *Aquacult. Eng.* **1999**, *20*, 261–276. [CrossRef]
33. Moutopoulos, D.K.; Stergiou, K.I. Length-weight and length-length relationships of fish species from Aegean Sea (Greece). *J. Appl. Ichthyol.* **2002**, *18*, 200–203. [CrossRef]
34. Vargas-Caro, C.; Bustamante, C.; Lamilla, J.; Bennett, M.B. A review of the longnose skates *Zearaja chilensis* and *Dipturus trachyderma* (Rajiformes: Rajidae). *Univ. Sci.* **2015**, *20*, 321–359. [CrossRef]
35. Yeldan, H.; Avsar, D.; Manasirh, M. Age, growth and feeding of the common stingray (*Dasyatis pastinaca*, L. 1758) in the Cilician coastal basin, northeastern Mediterranean Sea. *J. Appl. Ichthyol.* **2009**, *25*, 98–102. [CrossRef]
36. Yeldan, H.; Gundogdu, S. Morphometric relationships and growth of common stingray, *Dasyatis pastinaca* (Linnaeus, 1758) and marbled stingray, *Dasyatis marmorata* (Staindachner, 1892) in the northeastern Levantine Basin. *J. Black Sea Medit. Environ.* **2018**, *24*, 10–27.
37. Tiralongo, F.; Messina, G.; Cazzolla Gatti, R.; Tibullo, D.; Lombardo, B.M. Some biological aspects of juveniles of the rough ray, *Raja radula* Delaroche, 1809 in Eastern Sicily (central Mediterranean Sea). *J. Sea Res.* **2018**, *142*, 174–179. [CrossRef]
38. Tiralongo, F.; Messina, G.; Brundo, M.V.; Lombardo, B.M. Biological aspects of the common torpedo, *Torpedo torpedo* (Linnaeus, 1758) (Elasmobranchii: Torpedinidae), in the central Mediterranean Sea (Sicily, Ionian Sea). *Eur. Zool. J.* **2019**, *86*, 488–496. [CrossRef]
39. Romanelli, M.; Consalvo, I.; Vacchi, M.; Finoia, M.G. Diet of *Torpedo torpedo* and *Torpedo marmorata* in a coastal area of Central Western Italy (Mediterranean Sea). *Mar. Life* **2006**, *16*, 21–30.
40. Saglam, H.; Ak, O.; Kutlu, S.; Aydin, I. Diet and feeding strategy of the common stingray *Dasyatis pastinaca* (Linnaeus, 1758) on the Turkish coast of southeastern Black Sea. *Cah. Biol. Mar.* **2010**, *51*, 37–44. [CrossRef]
41. Santic, M.; Paladin, A.; Agovic, A. Diet of the common stingray, *Dasyatis pastinaca* (Chondrichthyes: Dasyatidae) in the eastern Adriatic Sea. *Cah. Biol. Mar.* **2011**, *52*, 349–356. [CrossRef]
42. Ponte, D.; Barcelos, L.D.M.; Santos, C.; Medeiros, J.; Barreiros, P. Diet of *Dasyatis pastinaca* and *Myliobatis aquila* (Myliobatiformes) from the Azores, NE Atlantic. *Cybium* **2016**, *40*, 209–214.
43. Compagno, L.J.V. Alternate life-history styles of cartilaginous fishes in time and space. *Environ. Biol. Fish.* **1990**, *28*, 33–75. [CrossRef]
44. Motta, P.J. Prey capture behavior and feeding mechanisms of elasmobranchs. In *Biology of Sharks and Their Relatives;* Carrier, J.C., Musick, J.A., Heithaus, M.R., Eds.; CRC Press: New York, NY, USA, 2004; pp. 165–202.
45. Bezerra, M.F.; Lacerda, L.D.; Lai, C.-T. Trace metals and persistent organic pollutants contamination in batoids (Chondrichthyes: Batoidea): A systematic review. *Environ. Pollut.* **2019**, *248*, 684–695. [CrossRef] [PubMed]
46. Peterson, C.H.; Fodrie, F.J.; Summerson, H.C.; Powers, S.P. Site-specific and density-dependent extinction of prey by schooling rays: Generation of a population sink in top quality habitat for bay scallops. *Oecologia* **2001**, *129*, 349–356. [CrossRef]
47. Thrush, S.F.; Pridmore, J.E.; Hewitt, J.E.; Cummings, V.J. The importance of predators on a sandflat: Interplay between seasonal changes in prey density and predator effects. *Mar. Ecol. Prog. Ser.* **1994**, *107*, 211–222. [CrossRef]
48. Yemisken, E.; Forero, M.G.; Megalofonou, P.; Eryilmaz, L.; Navarro, J. Feeding habits of three Batoids in the Levantine Sea (north-eastern Mediterranean Sea) based on stomach content and isotopic data. *J. Mar. Biol. Assoc. UK* **2017**, *98*, 89–96. [CrossRef]
49. Paterson, B.; Kainge, P. Rebuilding the Namibian hake fishery: A case for collaboration between scientists and fishermen. *Ecol. Soc.* **2014**, *19*, 49. [CrossRef]
50. Tiralongo, F.; Lillo, A.O.; Tibullo, D.; Tondo, E.; Lo Martire, C.; D'Agnese, R.; Macali, A.; Mancini, E.; Giovos, I.; Coco, S.; et al. Monitoring uncommon and non-indigenous fishes in Italian waters: One year of results for the AlienFish project. *Reg. Stud. Mar. Sci.* **2018**. [CrossRef]

Journal of
Marine Science and Engineering

MDPI

Review

Review of Estimating Trophic Relationships by Quantitative Fatty Acid Signature Analysis

Junbo Zhang [1,2,3,4,*], Chonglan Ren [1,†], Hu Zhang [5], Fang Yin [6], Shuo Zhang [1,*], Rong Wan [1,4,*] and Daisuke Kitazawa [7]

1 College of Marine Sciences, Shanghai Ocean University, Shanghai 201306, China; renchonglan_cz88@163.com
2 National Engineering Research Center for Oceanic Fisheries, Shanghai 201306, China
3 Key Laboratory of Sustainable Exploitation of Oceanic Fisheries Resources, Ministry of Education, Shanghai 201306, China
4 Key Laboratory of Oceanic Fisheries Exploration, Ministry of Agriculture and Rural Affairs, Shanghai 201306, China
5 Laboratory of Resource and Environment, Maine Fisheries Research Institution of Jiangsu, Nantong 226000, China; ahu80@163.com
6 College of Ocean Science and Engineering, Shanghai Maritime University, Shanghai 201306, China; fangyin@shmtu.edu.cn
7 Institute of Industrial Science, the University of Tokyo, 5-1-5 Kashiwanoha, Kashiwa-shi, Chiba 277-8574, Japan; dkita@iis.u-tokyo.ac.jp
* Correspondence: zhangjunbo1985@gmail.com (J.Z.); s-zhang@shou.edu.cn (S.Z.); rwan@shou.edu.cn (R.W.)
† The second author contributed equally to this work and is the co-first author.

Received: 26 October 2020; Accepted: 14 December 2020; Published: 18 December 2020

check for updates

Abstract: The dynamic predator–prey relations in the food web are vital for understanding the function and structure of ecosystems. Dietary estimation is a research hotspot of quantitative ecology, providing key insights into predator–prey relationships. One of the most promising approaches is quantitative fatty acid signature analysis (QFASA), which is the first generation of statistical tools to estimate the quantitative trophic predator–prey relationships by comparing the fatty acid (FA) signatures among predators and their prey. QFASA has been continuously widely applied, refined and extended since its introduction. This article reviewed the research progress of QFASA from development and application. QFASA reflects the long-term diet of predator, and provides the quantitative dietary composition of predator, but it is sensitive to the metabolism of predator. The calibration coefficients (CCs) and the FA subset are two crucial parameters to explain the metabolism of predators, but the incorrect construction or improper use of CCs and the FA subset may cause bias in dietary estimation. Further study and refinement of the QFASA approach is needed to identify recommendations for which CCs and subsets of FA work best for different taxa and systems.

Keywords: quantitative fatty acid signature analysis; aquatic food webs; dietary estimation

1. Introduction

To understand the structure and function of aquatic ecosystems and realize the trophic predator–prey relationships, it is necessary to obtain accurate dietary composition of predators. In aquatic ecosystems, the impracticality and limitations of direct observation method of feeding have prompted the development of indirect dietary estimation methods [1], such as the stomach content analysis [2], the stable isotopes (SI) analysis [3] and fatty acid (FA) analysis [4]. As one of the promising methods, FA analysis is based on the fact that specific FA can only be produced by certain primary producers (usually algae and bacteria), and higher trophic levels cannot synthesize by themselves; therefore, it can be used to track food sources. Fatty acids (FAs) have been widely applied in qualitative studies for analyzing the trophic predator–prey relationships. Then, Iverson et al. [5] developed a new method named quantitative fatty acid signature analysis (QFASA), which was the first time FAs have been used to quantitatively estimate the diet of predators.

The QFASA model has been continuously refined and extended, though it has been extensively applied in diet estimation. This article reviews the development of QFASA in the dietary estimation of aquatic species, and discusses the application scope of QFASA based on its merits and demerits. The aim of this paper is to provide a reference when applying QFASA to the quantitative evaluation of trophic relationship in marine ecosystem.

2. Fundamental Requirements of QFASA

QFASA is related to the characteristics of FAs. The differences in FA biosynthesis among species make it possible to identify the original source of certain FA. Firstly, FAs in prey are transferred, largely unaltered or at least in a predictable manner, to lipid storage sites (i.e., adipose tissue) of their predators, and reflect the long-term diet. Secondly, FAs derived from food are called dietary FAs, some of which can only be produced by specific species (such as algae and bacteria), and are essential for the survival of organisms, also called essential FAs [6]. Thirdly, in contrast to primary producers, other species usually can only produce some simple FAs (e.g., 14:0, 16:0, and 18:0 saturated FA; 14:1n-5, 16:1n-7, and 18:1n-9 monounsaturated FA) [7].

To understand the QFASA model, it is necessary to understand the terms of FA signature and calibration coefficient (CC).

The first issue is FA signature. According to Iverson et al. [5], the quantitative distribution of all FAs obtained in samples (prey or predators) is called FA signature. Each prey species has a unique pattern of FA signature due to the difference of life history and dietary habits [8]. As these signatures are conserved through the food chain, they can act as indicators of dietary composition [8].

The second issue is CC, which is a coefficient to calibrate the lipid metabolism process in the fat tissue of the predator [5]. Due to the biosynthesis, mobilization, modification and deposition of FAs, the FA signature of the predator is not completely consistent with its prey. Calibration coefficients (CCs) were determined previously from feeding experiments of individuals on a given long-term diet, with the assumption that after such long-term feeding, the adipose tissue stored in the predator will be similar to the given diet [5,9–11]. Then it can be calculated as the discrepancy or ratio, between adipose tissue and diet levels of each FA [5], as follows:

$$CC = \text{Predator}/\text{Prey} \quad (1)$$

Fundamental requirements for the application of QFASA are as follows: (i) a predator tissue properly sampled, stored, extracted and analyzed for FA signature; (ii) an appropriate sampling analysis of FA signature of potential prey species for a predator; (iii) the lipid metabolism and deposition of predator estimated, usually through captive feeding experiments; (iv) a quantitative model to minimize the distance of FA signatures between a predator and its prey [5]. More detailed information about the initial QFASA model can be found in the study of Iverson et al. [5], and further discussion of the model can be found in the article by Budge et al. [12].

QFASA is the first generation of statistical tools to evaluate the quantitative trophic predator–prey relationships by comparing the FA signatures among predators and their prey [5]. Taking into account the metabolism of the predator, the technique aims to find the combination of prey, based on their FA signatures, which yields a combined FA signature similar to that of the predator [5]. QFASA is developed based on two main hypotheses: the representative prey signature database contains all potential prey taxa and CCs are known and accurate [13].

QFASA can be used for individuals as well as food webs. Figure 1 is revised from Magnone et al. [14], who applied QFASA to generate a food web model in the aquatic environment to find out the trophic relationships among the species. The arrows point from all predators to their prey and the percentage represents the proportion of prey in the diet of predator.

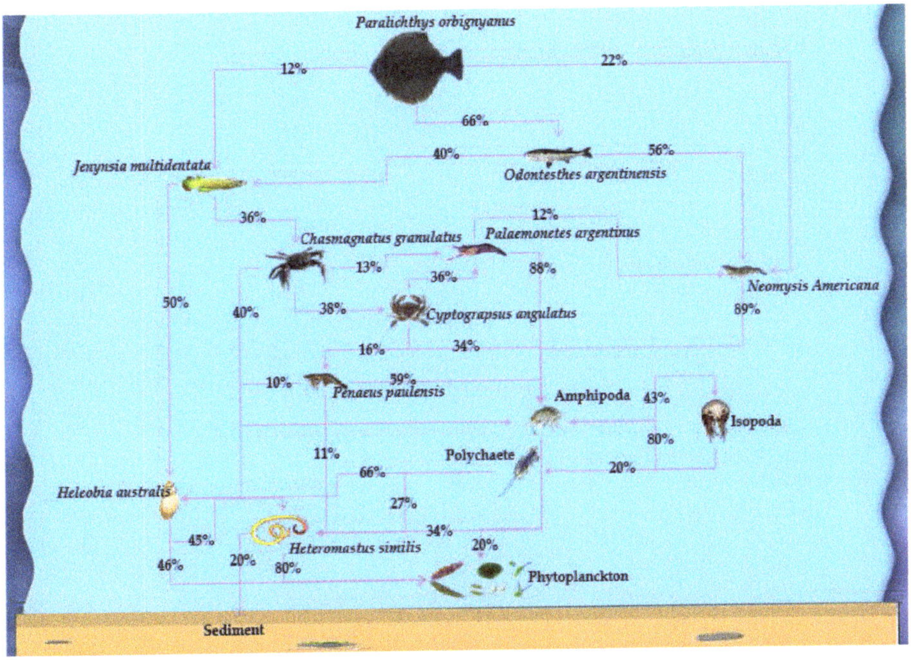

Figure 1. Quantitative trophic predator–prey relationships. (Revised from Magnone et al. [14]).

Ecological Modeling

The dietary assessment of each predator is acquired from the weighted combination of its prey's FA signatures, and then the weighted coefficient with the lowest statistical distance from the predator's FA signature is selected. This is achieved by minimizing the statistical distance between the prey's FA signatures weighted combination and predator's FA signature. The flow chart of QFASA application is shown in Figure 2 and the ecological modeling is as follows [5]:

The QFASA model considering the vectors of prey (x) and predator (y), originally developed by Iverson et al. [5], combined with the Kullback–Liebler (KL) distance, can be expressed as:

$$\hat{y} = \sum_k p_k \hat{x}_k \tag{2}$$

$$KL = \sum_j \left(y_j - \hat{y}_j\right) \log\left(\frac{y_j}{\hat{y}_j}\right) \tag{3}$$

where $\hat{y}$ is weighted sum of the FA of the prey, $\hat{x}_k$ is a vector of the prey species k, $\hat{x}_{kj}$ represents the mean of each FA j of the prey species k, p_k is a weight coefficient corresponding to the evaluated proportion of the k in the diet of predator, y_j represents the proportion of each FA j of the predator, and $\hat{y}_j$ represents the estimated value of each FA j of the predator.

The aim of the model is to select the p_k that makes the estimated value $\hat{y}$ as near as possible to the true value y. Because the p_k values are greater than or equal to zero and the sum of p_k values is 1, the evaluation can be transformed into a constrained optimization problem [14,15],

$$\min KL = \sum_j \left(y_j - \hat{y}_j\right) \log\left(\frac{y_j}{\hat{y}_j}\right) \quad (4)$$

$$\sum_k p_k = 1,\ p_k \geq 0 \quad (5)$$

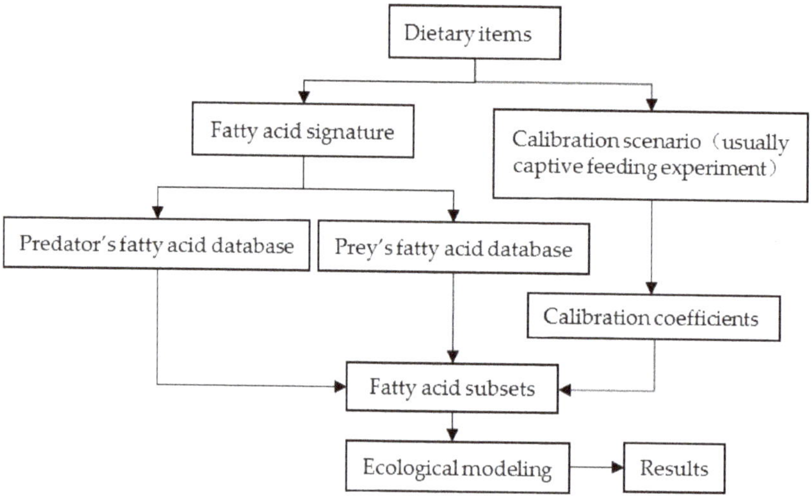

Figure 2. The application flow of quantitative fatty acid signature analysis (QFASA).

3. The Research Progress of QFASA

3.1. The Improvement of QFASA

The improvement of the QFASA model mainly includes the testing of the method in new taxa, selection and refinement of prey library, calculation of CC, selection of the FA subsets, and optimization of the statistical model.

3.1.1. The Testing of the Method in New Taxa

At first, QFASA was intended for estimating the dietary composition of marine mammals [5]. Subsequently, it has been used in multiple predators in various ecosystems, including marine mammals [16–19], seabirds [9–11,20], and fishes [14,15,21–23]. The development history of QFASA is shown in Figure 3, and each node (indicated by a circle) in the figure represents the first application to this species type.

In previous research, as shown in Figure 3, Iverson et al. [5] developed QFASA to estimate the dietary composition of *Halichoerus. grypus* and *Phoca. groenlandica*, and concluded that QFASA can be applicable to various predators and ecosystems; Iverson et al. [9] also applied QFASA to seabirds for the first time, and estimated the diets of *Uria. aalge* and red-legged *kittiwakes*; Wang et al. [10] estimated the dietary composition of threatened species (*Somateria. Wscheri* and *Polysticta. stelleri*) by QFASA, and inferred that QFASA can be used with other wild eiders or birds, and can estimate their diet composition at different life stages; Budge et al. [21] applied QFASA to marine fish for the first time and studied the diets of Atlantic salmon (*Salmo. salar*) (Figure 3); in addition, Magnone et al. [14] used

QFASA to generate a food web model in the aquatic environment to find out the trophic relationships among the species, and this was the first time that QFASA has been used to construct food web.

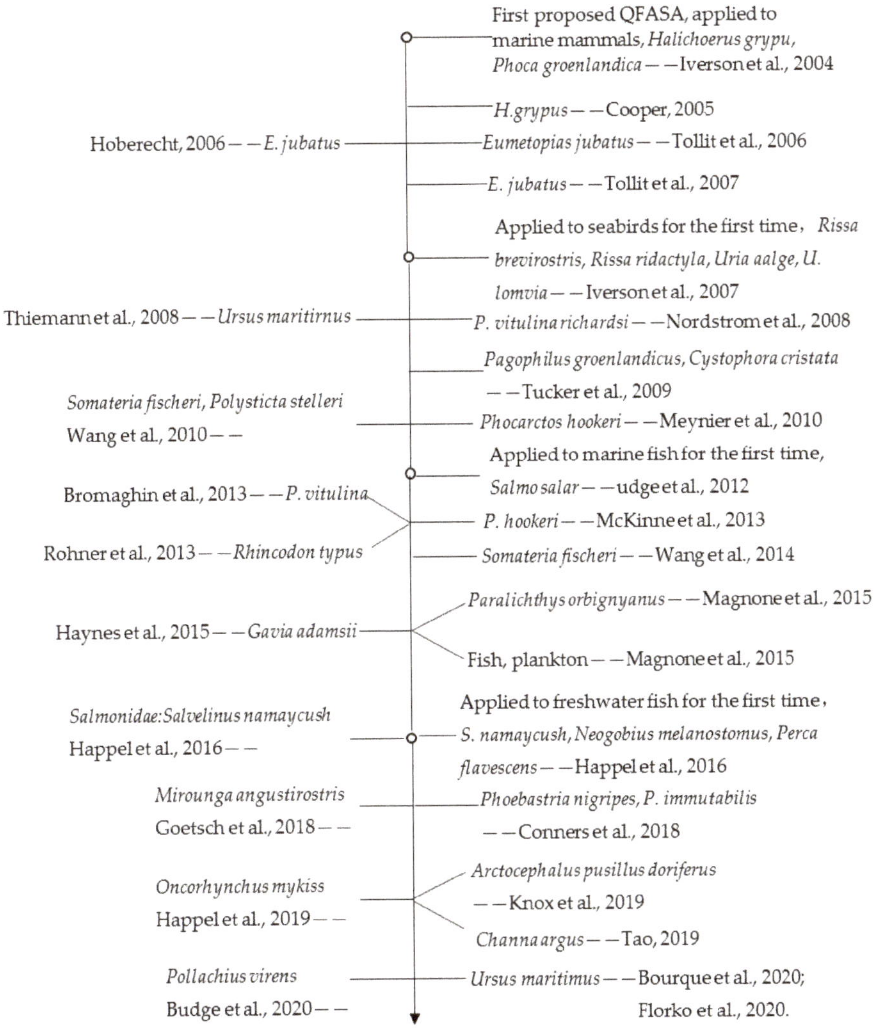

Figure 3. Development history of QFASA's dietary items.

In recent studies, Goetsch et al. [19] estimated the dietary composition of *Mirounga angustirostris* across five years; Conners et al. [20] studied the dietary composition of two albatross species through QFASA, which was beneficial to understand their fishery exploitation; Knox et al. [24] also evaluated the dietary composition of *Arctocephalus pusillus doriferus* by QFASA. It is obviously shown that the predator taxa of QFASA are dominated by marine species, mainly including marine mammals, seabirds, and fishes. In addition, as shown in Figure 3, there have been several studies focused on the diet of freshwater fishes in recent years [22,23,25,26], such as *Salvelinus namaycush* by Happel et al. [22].

3.1.2. The Selection and Refinement of Prey Library

A pre-condition of QFASA is a prey library which contains the potential prey species of predator [5], and each prey species has a unique FA signature. QFASA uses the average value of each prey to evaluate its contribution to the signature of the predator. However, wild animals consume individuals of different species rather than homogeneous species [5]. Although it is likely to identify the species in the ecosystem by its FA signature, there may be considerable differences within species, such as the same species but different ages or sizes. For instance, studies have found that the lipid content of sprat changes drastically with age and size [27]. Moreover, Magnone et al. [14] treated the young and adult *Paralichthys orbignyanus* as two types, and found that the adult *P. orbignyanus* was the top predator in the ecosystem studied while the young *P. orbignyanus* was prey of other species, and the dietary composition was different from each other. Therefore, species subgroups can be introduced into the QFASA model to provide extra details of diet. It is recommended that the age and size of prey should be taken into account when applying the QFASA model, and species with drastic changes in lipid content could be regarded as two types.

Besides, it is necessary to ensure that the quantity of FAs used in model is no less than the amount of potential prey in the database [19], otherwise, the estimated value of diet will be non-unique [18,28]. Three solutions are used to solve this problem based on previous studies. Firstly, multiple prey types with FAs alike could be assembled as a single type before applying QFASA [18,29,30]. Secondly, a post hoc method, assessing the diet of each prey as a unique type, is then pooling the estimated values of those prey into their respective communities [17]. Thirdly, Goetsch et al. [19] proposed an ad hoc approach (Drop Core Prey Analysis) to recognize the prey that did not contribute much to the diet and eliminate them from the model. In general, the first method may cause the mean FAs of the mixed prey species to be dissimilar to the practical prey, and the second method cannot ensure if the estimated value of the mixed diet is unique [19]. In contrast, the third method could avoid the deficiencies of the two methods above.

3.1.3. The Calculation of CCs

In addition to a detailed knowledge of the FA signatures of all potential prey, an understanding of the biochemistry and metabolism of FA within predators is essential to use this method accurately. At present, QFASA has two crucial parameters to explain the metabolism of predators: the set of CC for individual FA and the subset of FA used in modeling diets [5].

The first issue is CC, which accounts for the lipid metabolism of the predator. CCs were determined previously from captive feeding experiments. Although the CCs for a variety of birds, mammals, and fishes [11,15] have been determined, which is shown in Table 1, feeding experiments of most predators have not been conducted. In that condition, the developed CC of a surrogate species is normally used based on the evolutionary and ecological similarities. For instance, Thiemann et al. [16] used the CC of *Mustela vison* to replace the CC of polar bears; Meynier et al. [17] compared five sets of CC calculated by Iverson et al. [5] and Tollit et al. [31] then chose available CC that can be used for the application of QFASA on *Phocarctos hookeri*; Haynes et al. [30] used three CCs obtained by Iverson et al. [9] and Wang et al. [10] to replace the CC of *Gavia adamsii* by feeding trials.

Some studies have shown that diet estimation is very sensitive to the selection of CC, and the errors in CC would cause bias in dietary estimation [13,30]. Whereas, the factors such as feeding time and the type of food fed may affect the adaptability of the obtained CC in the study of wild predators [21,32] and the accuracy of CC in dietary estimation of wild animals cannot be verified. These issues have caused considerable criticism, and some recent studies emphasized and pointed out that using QFASA without CC is also sufficiently robust [15,21,22,32,33]. For instance, Budge et al. [21] used five calibration scenarios (four sets of CC were determined by captive feeding experiments, and the fifth calibration scenario was modeled without CC) to verify the application of QFASA in fish. They found a tendency to overestimate the dietary components used as prey for CC calculations, which indicates that applying QFASA without CCs to quantitatively estimate the dietary components and the trophic relationships in aquatic food web might be sufficiently robust. Magnone et al. [14] generated a food web model in the aquatic environment without CCs, found the results were similar to the previous knowledge; Happel et al. [22] illustrated that the prey used to make CCs can affect QFASA outputs and that perhaps ignoring CCs was a better method. Moreover, in a latest research, Bromaghin et al. [33] proposed a model that can simultaneously estimate CC and dietary composition on the basis of FA signature samples of predator and its preys alone. In summary, the application of CC depends on the species and system being studied, and the diet estimation of fish without CC is sufficiently robust, and for other taxa which are difficult to conduct captive feeding trials, it is recommended to refer to the model proposed by Bromaghin et al. [33].

Table 1. Captive feeding trials to improve the precision of quantitative fatty acid signature analysis (QFASA).

Species	Captive Feeding Trials	References
Halichoerus grypus, Phoca groenlandica	The CCs were originated from seals that fed herring for five months.	[5]
H. grypus	The CCs were obtained by replacing the mixed wild types with a homogeneous test diet via gastric cannula.	[8]
Eumetopias jubatus	The CCs were derived from the FA signature of blubber that has been fed only herring for a long time.	[34]
E. jubatus	Simulated ephemeral feeding events of *E. jubatus*.	[35]
Uria aalge	The CCs were obtained from silverside which had been raised in captivity since birth.	[9]
E. jubatus	Three CCs sets derived from constant diet respectively of *Clupea pallasii, Thaleichthys pacificus* or the mixture of them.	[31]
P. vitulina richardsi	The CCs were estimated from four seals fed herring.	[36]
Somateria fischeri, Polysticta stelleri	The female birds were grouped by type and kept in outdoor natural saline habitats respectively, and offered different proportions for each experimental diet, then proceed biopsy sampling.	[10]
Salmo salar	The changes of CCs with the FA signatures in the diet were investigated by feeding salmon with diet of different FA compositions.	[21]
S. fischeri	Developed CCs for individual FA for eggs.	[11]
Paralichthys orbignyanus	The CCs used for *P. orbignyanus* was obtained through the controlled feeding experiments of feeding on silverside and menhaden for 20 weeks.	[15]
Salmonidae: Salvelinus namaycush,	The juvenile lake trout was fed with invertebrates with diverse FA signatures to obtain CCs.	[22]
Mirounga angustirostris	The CCs were calculated from known dietary data derived from an adult female elephant seal in captivity.	[19]
Pollachius virens	The CCs were obtained by feeding *P. virens* on three formulated dietary.	[37]

3.1.4. The Selection of FA Subsets

The second issue, as stated above, is related to the FA subset used in assessing diets. More than 70 FAs have been identified through the gas chromatography and a polar capillary column in marine lipids studies [12]. Nevertheless, due to the influences of the predator's metabolisms, only some FAs can provide useful information for dietary estimation [38]. For example, short-chain or medium-chain FAs (i.e., <14 carbons) could only be produced by biosynthesis. In contrast, n-6 or n-3 polyunsaturated FAs usually only come from the diet. Due to these differences in origins, some studies have explored the utility of FA subsets rather than the full FAs.

Traditionally, the dietary or extended dietary subsets of FA investigated by Iverson et al. [5] is routinely used [10,11,13,16,30]. The former FAs are strictly from the diet, while the latter included several FAs that can be biosynthesized by the predator or ingested from diet [5]. In general, the extended dietary subset performed better in the modeling exercises, likely due to the increased information provided [8]. Besides, additional criteria were also applied by some investigators [18,19,21]. For example, Magnone et al. [15] selected seven FA sets to estimate diet; Budge et al. [21] used four FA sets; Goetsch et al. [19] adopted a new FA subset, which only included FAs derive from the diet (n = 46) and removed three FAs (16:4n-3, 18:1n-11, 22:1n-11) that were used in other studies. In conclusion, the optimum subsets of FA may depend on the type of predators (e.g., mammal, bird, fish) and the ecosystem studied (e.g., estuarine ecosystems, fresh water, and soil ecosystem). It is a significant area for further research.

There are generally three approaches to deal with the FA subset selected [13]: the "traditional" approach is to rescale the sum of the proportions of the FA components to 1 [5,13,19]; the "un-scaled" approach maintains the original ratio of FAs in the diet unchanged; and the "augmented" method is to add an additional ratio while maintaining the original proportions of FAs in the subset to make the sum equal to 1.

Bromaghin et al. [13] used computer simulations to compare the effects of three scaling methods based upon *KL* [5], Aitchison [39] and Chi square [39] distance measures on the variance and bias of dietary estimation. They recommended that researchers should check the sum of the original proportions and analyze the degree of difference among preys, and then determine whether the scaling will cause deviation.

3.1.5. The Optimization of the Statistical Model

QFASA has been extensively used since it was proposed and some distance measurement methods have been applied in previous studies (e.g., [5,19,39,40]), such as the *KL* [5], Aitchison [19,39], Chi square [39] statistical distance, although the *KL* measure originally recommended by Iverson et al. [5] has been frequently used [39].

Dietary estimation was usually carried out in two spaces. One was the prey space, which was transformed by dividing predator FA signature by CC based on Formula (1) [5]; and the other was the predator space, which was obtained by multiplying prey FA signature by CC based on Formula (1) [18,19]. Bromaghin et al. [41] investigated the influence of the choice of distance measures (such as *KL* or Aitchison) and optimization spaces (i.e., prey or predator space mentioned above) on dietary estimation in the application of the QFASA model, and revealed that the choice of evaluation method may have a significant impact on dietary estimation. Additionally, Bromaghin et al. [42] developed a new algorithm that can objectively determine the bootstrap sample size to generate a pseudo-predator signature with actual attributes, thus improving the efficacy of computer modeling to evaluate the performance of the QFASA estimator. Besides, Bromaghin et al. [40] used computer modeling to study the robustness of the Aitchison and *KL* distance measure, kept a record of the deviations in dietary estimation, and concluded that the former was more robust in terms of CC errors while the latter was more robust to the ingestion of preys unrepresentative in the prey profiles database. Furthermore, Litmanen et al. [43] estimated the performance of several algorithms and found that some Bayesian algorithms take longer time to calculate than QFASA, recommending the use of Chi square or *KL* statistical distance. In conclusion, the distance measure that performs the best may depend on the ecosystem being studied, and it is recommended to test multiple measures to assess differences, or evaluate the ecosystem under controlled conditions (e.g., feeding experiments).

In the application of the QFASA model, the statistical characteristics of predator dietary estimation is usually estimated through computer simulations [42], e.g., commercial software, Fortran programs [18], R package [13,19,30,32,44,45], a combination program of R and Fortran [6,41], and Matlab with its optimization toolbox [14,15,17,36]. In recent years, Bromaghin summarized a new R package named QFASAR, calculating the goodness-of-fit diagnosis, which may enhance the performance of the prey signature database [19,44].

3.2. Research Status of QFASA

QFASA has become a promising approach in dietary estimation since it was proposed and its applications are shown in Table 2. In recent years, Wang et al. [11] estimated the diets of threatened *S. fischeri*, and concluded that infertile eggs yolks can be used to evaluate the diet of breeding female *S. fischeri* for better understanding the source and timing of nutrients during reproduction; Magnone et al. [15] estimated the diet of *P. orbignyanus* by QFASA, determined CC and validated the model with a controlled experiment, which quantified the diet of lower vertebrates (*P. orbignyanus*) for the first time; Magnone et al. [14] used QFASA to generate a food web model in the aquatic environment to find out the trophic relationships among the species; Conners et al. [20] estimated two albatross species by QFASA, and adapted the QFASA model, which introduced a combination of FAs and fatty alcohols; Goetsch et al. [19] estimated the diet composition of *M. angustirostris*, and suggested that the seals mainly fed on mesopelagic fishes; Knox et al. [24] estimated the diet of *A. pusillus doriferus*, finding that elasmobranchs accounted for more in the diet of males than previous reports, and showed that prey composition varied among males; Happel et al. [25] used a controlled trial to test the influence of intraspecific difference in FA signatures of prey on the QFASA model, indicating that QFASA used for steelhead trout may not only be used for a specific lake, but also for other freshwater systems with alewife and round goby as the main food; Tao [26] estimated the diet composition of *Channa. argus* using the QFASA model, and then calculated the biomagnifcation factors (BMFs) of the alternative halogenated flame retardants (AHFRs) based upon the results.

Table 2. The application of QFASA.

Species	Species Type	Research Contents	References
H. grypus, *P. groenlandica*	Marine mammals	Proposed the QFASA model, which was the first time FAs has been used to quantitatively estimate the diet of predators.	[5]
H. grypus	Marine mammals	Conducted controlled feeding experiments designed to investigate some of the issues relating to CC, optimization of FA sets and validation of QFASA.	[8]
E. jubatus	Marine mammals	Captive feeding experiments were conducted on juvenile female *E. jubatus* to estimate the ability of QFASA to recognize known mixed dietary.	[34]
E. jubatus	Marine mammals	Simulated ephemeral feeding events of *E. jubatus* and investigated the use of QFASA to evaluate diet.	[35]
Rissa brevirostris, *Rissa ridactyla*, *U. aalge*, *U. lomvia*	Seabirds	Using four species of captive and wild birds to verify the application of QFASA in seabird dietary estimation.	[9]
P. vitulina richardsi	Marine mammals	Tested the QFASA model with newly weaned *P. vitulina richardsi*, obtained species-specific CCs, and compared the credibility of QFASA with previous studies.	[36]
Ursus maritirnus	Marine mammals	Examined the internal and external factors that affect the prey selection of *U. maritirnus*.	[16]
Pagophilus groenlandicus, *Cystophora cristata*	Marine mammals	Studied the effects of demographic, time and space differences on the diet of *P. groenlandicus* and *C. cristata*.	[46]
Phocarctos hookeri	Marine mammals	Optimized the QFASA model for bycaught *P. hookeri* by changing some parameters (such as the CC subsets used in previous studies [5,17,30]); and estimated the diet of *P. hookeri* by optimization model.	[17]

Table 2. *Cont.*

Species	Species Type	Research Contents	References
S. fischeri, *P. stelleri*	Seabirds	Developed CCs to explain the lipid metabolism of eider and verified the reliability of QFASA through captive experiments.	[10]
S. salar	Marine fish	Investigated the application of QFASA to *S. salar*.	[21]
Ursus maritimus	Marine mammals	Used QFASA and δ^{13}C-FA to estimate the variations in the diet of *U. maritimus* from 1984 to 2011, which reflected the massive shifts in the ecosystem of East Greenland.	[47]
P. vitulina	Marine mammals	Used QFASA to investigate the diets of *P. vitulina*.	[18]
Rhincodon typus	Marine fish	Used QFASA and stomach content analysis to identify the main prey of *R. typus*.	[48]
S. fischeri	Seabirds	Examined the diversity of FA signatures between fertilized and unfertilized eggs of feeding female *S. fischeri*, developed CC for individual to explain the modification of yolk on FAs derived from the dietary, and verified that the QFASA model could use yolk to estimate the diet of maternal.	[11]
P. orbignyanus	Marine fish	Estimated the quantitative diet of *P. orbignyanus*, determined CC and validated the model by controlled trials.	[15]
P. orbignyanus and other 21 species	Marine fish, plankton etc.	Used QFASA to generate a food web model in the aquatic environment to find out the trophic relationships among the species and compared the consequences with previous researches.	[14]
Gavia adamsii	Seabirds	Estimated the diet of *G. adamsii* breeding on the Arctic Coastal Plain using QFASA.	[30]
Salmonidae: S. namaycush	Marine mammals	Determined the time it takes to make the FA signatures of the *S. namaycush* similar to its dietary, and evaluated the difference in CC among different diets.	[22]
S. namaycush, *Neogobius melanostomus*, *Perca flavescens*	Marine mammals	Established a model for estimating diet and evaluated its ability in freshwater predator.	[23]
Phoebastria nigripes, *Phoebastria immutabilis*	Seabirds	Estimated two albatross types by QFASA, and adapted the QFASA model, which introduced a combination of FAs and fatty alcohols.	[20]
M. angustirostris	Marine mammals	Proposed a new approach to optimize prey library, then evaluated the diet of *M. angustirostris* with the optimized prey library.	[19]
Arctocephalus pusillus doriferus	Marine mammals	Investigated the diet composition of male *A. pusillus doriferus*, and dietary differences among males and seasonal changes in diet.	[24]
Oncorhynchus mykiss	Freshwater fish	The effect of intraspecific difference in the FASs of preys on QFASA was verified by controlled captive trial.	[25]
Channa argus	Freshwater fish	Estimated the diet composition of *C. argus* using QFASA model.	[26]
P. virens	Marine fish	Estimated the effects of diet fat concentrations on assimilation patterns of FAs in *P. virens*.	[37]
Ursus maritimus	Marine mammals	Estimated the diet composition of *U. maritimus* by QFASA.	[49,50]

4. Conclusions and Suggestions

FAs have been widely applied in qualitative research for evaluating trophic predator–prey relationships. However, QFASA is the first attempt to use FAs to quantitatively estimate the diet of predators, which provides an insight into the function and structure of dynamic ecosystems. The core of QFASA is to find the FA signatures combination of prey that is most similar to the FA signatures of predator to infer the diet of predator. One of the most important assumptions of the QFASA model is that the metabolism of the predator is already known. Currently, the effect of predators on FA metabolism and deposition is explained by CCs, which were determined previously from captive feeding trials and could be found by calculating FA levels found in the predator over FA levels in the food. Additionally, CCs have already been evaluated in several species of marine mammals, seabirds and fish (Table 1). The refinement of which subsets of FAs are more effective for QFASA accuracy remains a topic of study. QFASA has been used in a variety of taxa (e.g., marine mammals, seabirds and fish) rather than being limited to high-trophic vertebrates, and presumably it will be tested in new taxa. If one predator in a given ecosystem meets all of the above requirements for using QFASA, it can be modeled.

Nevertheless, QFASA has several limitations. Firstly, since it is a non-probabilistic model, it is hard to evaluate the effects of the different uncertainties related with dietary proportion estimation. The handling of confidence interval in QFASA can be found in Stewart et al. [51,52]. Ecological mechanisms cannot be built directly into the model owing to the absence of an explicit model [53]. Secondly, some species are not suitable for captive experiments; therefore, accurate CCs cannot be obtained. Thirdly, the model is not applicable to all species, because some taxa lack specific fat storage organs. In addition, some factors may cause bias when QFASA is applied to fish and invertebrates, since FAs may be modified during metabolism and transportation. Some invertebrates tend to have greater ability to biosynthesize and modify the FA than higher trophic organisms, and fish have a stronger ability to modify some exogenous FAs than mammals and birds [7]; many studies have indicated that FAs with a high proportion in the diets of fish are easily catabolized [21]; the growth and reproductive stage (e.g., the proportions of 20:5n-3 and 22:6n-3 in fish eggs seems to be fixed) may affect FA signature of tissues [21]; some freshwater taxa (e.g., Daphnia) can elongate and desaturate 18:3n-3 to 20:5n-3 [7].

Several techniques have been used to study predator–prey relationships, such as stomach content analysis, IS analysis and FA analysis [54]. Traditional stomach content analysis can directly reflect the diet composition of predator, but it usually only represents a snapshot of diet and tends to underestimate the proportion of soft-bodied prey [55]. Chemical markers such as SI and FAs can reflect the long-term diet of predator. SI analysis plays an important role in evaluating the trophic level of predators, but the resolution is limited by the number of SI to be measured (e.g., typically only 2–3 SI are measured) [56]. FAs seem to provide more information than SI, because many potential FAs can be measured. However, the proportion of the measured FAs is always limited to a sum of one. Therefore, some studies have developed mixed models that combine SI and FAs, such as the Bayesian fatty acid-based mixing model (Fatty Acid Source Tracking Algorithm in R, FASTAR) [57,58] intended for zooplankton and benthic macro-invertebrates, a mixed model combining FAs with SI (FastinR), which could improve the dietary estimation with the available fat content and conversion coefficients [53], and the Bayesian mixing model framework (mixSIAR) [59], which evaluates the relative contribution of food sources in the diet of predator. As a promising method in quantitative ecosystem, QFASA relies on the distance measurement rather than model-based formulation to evaluate the most likely diet proportions, which is contrary to (Bayesian) SI and FA mixing models.

As for the future studies of QFASA, firstly, it is suggested to use QFASA in combination with other approaches (such as the SI analysis [8] or stomach content analysis [16]) to obtain more comprehensive information. Secondly, some simplification procedures of FAs extraction are expected to widen the use of the QFASA model in marine ecology and biology, such as the simplified method to extract polyunsaturated FAs [60]. Thirdly, the application of QFASA in invertebrates and low-trophic vertebrates needs further study. More studies are expected to use the QFASA model to build food webs [14,61]. For instance, Magnone et al. [14] generated a food web model in the aquatic environment to find out the trophic relationships among the species by QFASA. Further, captive feeding trials may provide a good way of testing if QFASA will work and its intricacies, but the predator signature produced through trials may not relate to wild species of different age, size, or lipid status even if they are fed the same prey in the wild. There is still much remaining to be known and tested with how FAs can be used in quantitative models. For instance, Goetz et al. [62] indicated that within the lake trout species different ectomorphs accumulate lipids in different ways which may mean finding CCs for plastic species may be difficult and suggests that over time CCs would need to be updated as genetics drift.

In conclusion, it has been clearly shown that FA has become an important trophic tracer in the research of carbon transfer, predator–prey relationships, food webs, along with the function and structure of dynamic ecosystem. Although QFASA has some limitations, it is still the promising approach because of its several potential advantages [12]. QFASA reflects the long-term (a period of weeks to months) diet compared with the stomach content analysis, allows more than three prey types to be estimated, and can avoid the problem of underdetermined systems common in analysis with SI markers. Results of QFASA may be useful in studies into the effects of maternal diets on offspring or the assessment of nutrient deficiencies in marine organisms [63,64].

Author Contributions: Conceptualization, J.Z.; methodology, J.Z. and C.R.; data analysis, C.R., H.Z., F.Y. and S.Z.; resources, H.Z. and S.Z.; writing—original draft preparation, J.Z. and C.R.; writing—review and editing, J.Z., C.R. and D.K.; supervision, R.W. and D.K. All authors have read and agreed to the published version of the manuscript.

Funding: This study is partly supported by the National Key Research and Development Program of China (No. 2019YFC0312104), the Young Orient Scholars Program of Shanghai (No. QD2017038), and the National Natural Science Foundation of China (No. 41807341).

Acknowledgments: The authors would like to thank Yang Chenxing for discussing the English grammar of the manuscript.

Conflicts of Interest: The authors declare no conflict of interest. The funders had no role in the design of the study; in the collection, analyses, or interpretation of data; in the writing of the manuscript, or in the decision to publish the results.

References

1. Kelly, J.R.; Scheibling, R.E. Fatty acids as dietary tracers in benthic food webs. *Mar. Ecol. Prog. Ser.* **2012**, *446*, 1–22. [CrossRef]
2. Scott, T. Some further observations on the food of fishes, with a note on the food observed in the stomach of a common porpoise. *Annu. Rep. Fish. Board Scotl. Sci. Investig.* **1903**, *21*, 218–227.
3. Fry, B.; Sherr, E.B. δ13C measurements as indicators of carbon flow in marine and freshwater ecosystems. In *Stable Isotopes in Ecological Research*, 2nd ed.; Springer: New York, NY, USA, 1989; pp. 196–229.
4. Sargent, J.R.; Falkpetersen, S. The lipid biochemistry of calanoid copepods. *Hydrobiologia* **1988**, *167*, 101–114. [CrossRef]
5. Iverson, S.J.; Field, C.; Bowen, W.D.; Blanchard, W. Quantitative fatty acid signature analysis: A new method of estimating predator diets. *Ecol. Monogr.* **2004**, *74*, 211–235. [CrossRef]
6. Olsen, Y. Lipids and essential fatty acids in aquatic food webs: What can freshwater ecologists learn from mariculture. In *Lipids in Freshwater Ecosystems*, 2nd ed.; Springer: New York, NY, USA, 1999; Volume 8, pp. 161–202.
7. Iverson, S.J. Tracing aquatic food webs using fatty acids: From qualitative indicators to quantitative determination. In *Lipids in Aquatic Ecosystems*, 2nd ed.; Springer: New York, NY, USA, 2009; Volume 12, pp. 281–308.
8. Cooper, M.H. Fatty Acid Metabolism in Marine Carnivores: Implications for Quantitative Estimation of Predator Diets. Ph.D. Thesis, Dalhousie University, Halifax, NS, Canada, 2004.
9. Iverson, S.J.; Springer, A.M.; Kitaysky, A.S. Seabirds as indicators of food web structure and ecosystem variability: Qualitative and quantitative diet analyses using fatty acids. *Mar. Ecol. Prog. Ser.* **2007**, *352*, 235–244. [CrossRef]

10. Wang, S.W.; Hollmén, T.E.; Iverson, S.J. Validating quantitative fatty acid signature analysis to estimate diets of spectacled and Steller's eiders (*Somateria fischeri* and *Polysticta stelleri*). *J. Comp. Physiol. B Biochem. Syst. Environ. Physiol.* **2010**, *180*, 125–139. [CrossRef]
11. Wang, S.W.; Hollmén, T.E.; Iverson, S.J. Egg yolk fatty acids as a proxy to quantify diets of female spectacled eiders (*Somateria fischeri*). *Can. J. Zool.* **2014**, *92*, 453–461. [CrossRef]
12. Budge, S.M.; Iverson, S.J.; Koopman, H.N. Studying trophic ecology in marine ecosystems using fatty acids: A primer on analysis and interpretation. *Mar. Mammal Sci.* **2006**, *22*, 759–801. [CrossRef]
13. Bromaghin, J.F.; Budge, S.M.; Thiemann, G.W. Should fatty acid signature proportions sum to 1 for diet estimation? *Ecol. Res.* **2016**, *31*, 597–606. [CrossRef]
14. Magnone, L.; Bessonart, M.; Gadea, J.; Salhi, M. Trophic relationships in an estuarine environment: A quantitative fatty acid analysis signature approach. *Estuar. Coast. Shelf Sci.* **2015**, *166*, 24–33. [CrossRef]
15. Magnone, L.; Bessonart, M.; Rocamora, M.; Gadea, J.; Salhi, M. Diet estimation of *Paralichthys orbignyanus* in a coastal lagoon via quantitative fatty acid signature analysis. *J. Exp. Mar. Biol. Ecol.* **2015**, *462*, 36–49. [CrossRef]
16. Thiemann, G.W.; Iverson, S.J.; Stirling, I. Polar bear diets and arctic marine food webs: Insights from fatty acid analysis. *Ecol. Monogr.* **2008**, *78*, 591–613. [CrossRef]
17. Meynier, L.; Morel, P.C.H.; Chilvers, B.L.; Mackenzie, D.D.S.; Duignan, P.J. Quantitative fatty acid signature analysis on New Zealand sea lions: Model sensitivity and diet estimates. *J. Mammal* **2010**, *91*, 1484–1495. [CrossRef]
18. Bromaghin, J.F.; Lance, M.M.; Elliott, E.W.; Jeffries, S.J.; Acevedo-Gutiérrez, A.; Kennish, J.M. New insights into the diets of harbor seals (*Phoca vitulina*) in the Salish Sea revealed by analysis of fatty acid signatures. *Fish. Bull.* **2013**, *111*, 13–26. [CrossRef]
19. Goetsch, C.; Conners, M.G.; Budge, S.M.; Mitani, Y.; Walker, W.A.; Bromaghin, J.F. Energy-Rich mesopelagic fishes revealed as a critical prey resource for a deep-diving predator using quantitative fatty acid signature analysis. *Front. Mar. Sci.* **2018**, *5*, 1–19. [CrossRef]
20. Conners, M.G.; Chandra, G.; Budge, S.M.; Walker, W.A.; Yoko, M.; Costa, D.P. Fisheries Exploitation by Albatross Quantified With Lipid Analysis. *Front. Mar. Sci.* **2018**, *5*, 1–17. [CrossRef]
21. Budge, S.M.; Penney, S.N.; Lall, S.P.; Trudel, M. Estimating diets of Atlantic salmon (*Salmo salar*) using fatty acid signature analyses; validation with controlled feeding studies. *Can. J. Fish. Aquat. Sci.* **2012**, *69*, 1033–1046. [CrossRef]
22. Happel, A.; Stratton, L.; Kolb, C.; Hays, C.; Rinchard, J.; Czesny, S. Evaluating quantitative fatty acid signature analysis (QFASA) in fish using controlled feeding experiments. *Can. J. Fish. Aquat. Sci.* **2016**, *73*, 1222–1229. [CrossRef]
23. Happel, A.; Stratton, L.; Pattridge, R.; Rinchard, J.; Czesny, S. Fatty-acid profiles of juvenile lake trout reflect experimental diets consisting of natural prey. *Freshw. Biol.* **2016**, *61*, 1466–1476. [CrossRef]
24. Knox, T.C.; Callahan, D.L.; Kernaléguen, L.; Baylis, A.M.M.; Arnould, J.P.Y. Blubber fatty acids reveal variation in the diet of male Australian fur seals. *Mar. Biol.* **2019**, *166*, 117. [CrossRef]
25. Happel, A.; Maier, C.; Farese, N.V.; Czesny, S.; Rinchard, J. Fatty acids differentiate consumers despite variation within prey fatty acid profiles. *Freshw. Biol.* **2019**, *64*, 1416–1426. [CrossRef]
26. Tao, L. Bioaccumulation Characteristics and Trophic Magnification of Alternative Halogenated Flame Retardants. Ph.D. Thesis, University of Chinese Academy of Sciences (Guangzhou Institute of Geochemistry Chinese Academy of Sciences), Guangzhou, China, 2019.
27. Keinänen, M.; Käkelä, R.; Ritvanen, T.; Pönni, J.; Harjunpää, H.; Myllylä, T.; Vuorinen, P.J. Fatty acid signatures connect thiamine deficiency with the diet of the Atlantic salmon (*Salmo salar*) feeding in the Baltic Sea. *Mar. Biol.* **2018**, *165*, 161. [CrossRef] [PubMed]
28. Phillips, D.L.; Inger, R.; Bearhop, S.; Jackson, A.L.; Moore, J.W.; Parnell, A.C. Best practices for use of stable isotope mixing models in food-web studies. *Can. J. Zool.* **2014**, *92*, 823–835. [CrossRef]
29. Piché, J.; Iverson, S.J.; Parrish, F.A.; Dollar, R. Characterization of forage fish and invertebrates in the Northwestern Hawaiian Islands using fatty acid signatures: Species and ecological groups. *Mar. Ecol. Prog. Ser.* **2010**, *418*, 1–15. [CrossRef]
30. Haynes, T.B.; Schmut, J.A.; Bromaghin, J.F.; Iverson, S.J.; Padula, V.M.; Rosenberger, A.E. Diet of yellow-billed loons (*Gavia adamsii*) in Arctic lakes during the nesting season inferred from fatty acid analysis. *Polar Biol.* **2015**, *38*, 1239–1247. [CrossRef]

31. Tollit, D.J.; Rosen, D.A.S.; Iverson, S.J.; Trites, A.W. Stellar sea lion QFASA captive feeding studies. National Oceanic and Atmospheric Administration. *Final Rep. North Pac. Univ. Mar. Mammal Res. Consort.* **2007**, *9*, 1–35.
32. Rosen, D.; Tollit, D. Effects of phylogeny and prey type on fatty acid calibration coefficients in three pinniped species: Implications for the QFASA dietary quantification technique. *Mar. Ecol. Prog.* **2012**, *467*, 263–276. [CrossRef]
33. Bromaghin, J.F.; Budge, S.M.; Thiemann, G.W.; Rode, K.D. Simultaneous estimation of diet composition and calibration coefficients with fatty acid signature data. *Ecol. Evol.* **2017**, *7*, 6103–6113. [CrossRef]
34. Tollit, D.J.; Heaslip, S.G.; Deagle, B.E.; Iverson, S.I.; Joy, R.; Rosen, D.A.S.; Trites, A.W. Estimating diet composition in sea lions: Which technique to choose. In *Sea Lions of the World*; Trites, A.W., Atkinson, S.K., DeMaster, D.P., Fritz, L.W., Gelatt, T.S., Rea, L.D., Wynne, K.M., Eds.; Alaska Sea Grant College Program, University of Alaska Fairbanks: Fairbanks, AL, USA, 2006; pp. 293–308.
35. Hoberecht, L.K. Investigating the Use of Blubber Fatty Acids to Detect Steller Sealion (*Eumetopias jubatus*) Foraging on Ephemeral High-Quality Prey. Ph.D. Thesis, University of Washington, Seattle, WA, USA, 2006.
36. Nordstrom, C.A.; Wilson, L.J.; Iverson, S.J.; Tollit, D.J. Evaluating quantitative fatty acid signature analysis (QFASA) using harbour seals *Phoca vitulina richardsi* in captive feeding studies. *Mar. Ecol. Prog. Ser.* **2008**, *360*, 245–263. [CrossRef]
37. Budge, S.M.; Townsend, K.; Lall, S.P.; Bromaghin, J.F. Dietary fat concentrations influence fatty acid assimilation patterns in Atlantic pollock (*Pollachius virens*). *Philos. Trans. R. Soc. B* **2020**, *375*, 20190649. [CrossRef]
38. Iverson, S.J.; Bowen, W.D.; Boness, D.J.; Oftedal, O.T. The effect of maternal size and milk energy output on pup growth in grey seals (*Halichoerus grypus*). *Physiol. Biochem. Zool.* **1993**, *66*, 61–88. [CrossRef]
39. Stewart, C.; Iverson, S.J.; Field, C. Testing for a change in diet using fatty acid signatures. *Environ. Ecol. Stat.* **2014**, *21*, 775–792. [CrossRef]
40. Bromaghin, J.F.; Budge, S.M.; Thiemann, G.W.; Rode, K.D. Assessing the robustness of quantitative fatty acid signature analysis to assumption violations. *Methods Ecol. Evol.* **2016**, *7*, 51–59. [CrossRef]
41. Bromaghin, J.F.; Rode, K.D.; Budge, S.M.; Thiemann, G.W. Distance measures and optimization spaces in quantitative fatty acid signature analysis. *Ecol. Evol.* **2015**, *5*, 1249–1262. [CrossRef] [PubMed]
42. Bromaghin, J.F. Simulating realistic predator signatures in quantitative fatty acid signature analysis. *Ecol. Inform.* **2015**, *30*, 68–71. [CrossRef]
43. Litmanen, J.J.; Perälä, T.A.; Taipale, S.J. Comparison of Bayesian and numerical optimization-based diet estimation on herbivorous zooplankton. *Philos. Trans. R. Soc. B* **2020**, *375*, 20190651. [CrossRef] [PubMed]
44. Bromaghin, J.F. QFASAR: Quantitative fatty acid signature analysis with R. *Methods Ecol. Evol.* **2017**, *8*, 1158–1162. [CrossRef]
45. Bromaghin, J.F.; Budge, S.M.; Thiemann, G.W. Detect and exploit hidden structure in fatty acid signature data. *Ecosphere* **2017**, *8*, 1–13. [CrossRef]
46. Tucker, S.; Bowen, W.D.; Iverson, S.J.; Blanchard, W.; Stenson, G.B. Sources of variation in diets of harp and hooded seals estimated from quantitative fatty acid signature analysis (QFASA). *Mar. Ecol. Prog. Ser.* **2009**, *384*, 287–302. [CrossRef]
47. Mckinney, M.A.; Iverson, S.J.; Fisk, A.T.; Sonne, C.; Rigét, F.F.; Letcher, R.J. Global change effects on the long-term feeding ecology and contaminant exposures of East Greenland polar bears. *Glob. Chang. Biol.* **2013**, *19*, 2360–2372. [CrossRef]
48. Rohner, C.A.; Couturier, L.I.E.; Richardson, A.J.; Pierce, S.J.; Prebble, C.E.M.; Gibbons, M.J.; Nichols, P.D. Diet of whale sharks *Rhincodon typus* inferred from stomach content and signature fatty acid analyses. *Mar. Ecol. Prog. Ser.* **2013**, *493*, 219–235. [CrossRef]
49. Bourque, J.; Atwood, T.C.; Divoky, G.J.; Stewart, C.; McKinney, M.A. Fatty acid-based diet estimates suggest ringed seal remain the main prey of southern Beaufort Sea polar bears despite recent use of onshore food resources. *Ecol. Evol.* **2020**, *10*, 2093–2103. [CrossRef] [PubMed]
50. Florko, K.R.N.; Thiemann, G.W.; Bromaghin, J.F. Drivers and consequences of apex predator diet composition in the Canadian Beaufort Sea. *Oecologia* **2020**, *19*, 51–63. [CrossRef]
51. Stewart, C.; Field, C. Managing the essential zeros in quantitative fatty acid signature analysis. *J. Agric. Biol. Environ. Stat.* **2011**, *16*, 45–69. [CrossRef]

52. Stewart, C. Zero-inflated beta distribution for modeling the proportions in quantitative fatty acid signature analysis. *J. Appl. Stat.* **2013**, *40*, 985–992. [CrossRef]
53. Neubauer, P.; Jensen, O.P. Bayesian estimation of predator diet composition from fatty acids and stable isotopes. *PeerJ* **2015**, *3*, 1–19. [CrossRef] [PubMed]
54. Bowen, W.D.; Iverson, S.J. Methods of estimating marine mammal diets: A review of validation experiments and sources of bias and uncertainty. *Mar. Mammal Sci.* **2013**, *29*, 719–754. [CrossRef]
55. Brush, J.M.; Fisk, A.T.; Hussey, N.E.; Johnson, T.B. Spatial and seasonal variability in the diet of round goby (*Neogobius melanostomus*): Stable isotopes indicate that stomach contents overestimate the importance of dreissenids. *Can. J. Fish. Aquat. Sci.* **2012**, *69*, 573–586. [CrossRef]
56. Phillips, D.L. Mixing models in analyses of diet using multiple stable isotopes: A critique. *Oecologia* **2001**, *127*, 166–170. [CrossRef]
57. Galloway, A.W.E.; Eisenlord, M.E.; Dethier, M.N.; Holtgrieve, G.; Brett, M.T. Quantitative estimates of resource utilization by an herbivorous isopod using a Bayesian fatty acid mixing model. *Mar. Ecol. Prog. Ser.* **2014**, *507*, 219–232. [CrossRef]
58. Galloway, A.W.E.; Taipale, S.J.; Hiltunen, M.; Peltomaa, E.; Strandberg, U.; Brett, M.T.; Kankaala, P. Diet specific biomarkers show that high quality phytoplankton fuel herbivorous zooplankton in large boreal lakes. *Freshw. Biol.* **2014**, *59*, 1902–1915. [CrossRef]
59. Jankowska, E.; Troch, M.D.; Michel, L.N.; Lepoint, G.; Wodarska-Kowalczuk, M. Modification of benthic food web structure by recovering seagrass meadows, as revealed by trophic markers and mixing models. *Ecol. Indic.* **2018**, *90*, 28–37. [CrossRef]
60. Kang, J.X.; Wang, J.A. Simplified method for analysis of polyunsaturated fatty acids. *BMC Biochem.* **2005**, *6*, 5. [CrossRef] [PubMed]
61. Kühn, J.; Tobias, K.; Jähngen, A.; Ruess, L. Shifting systems: Prerequisites for the application of quantitative fatty acid signature analysis in soil food webs. *Philos. Trans. R. Soc. B* **2020**, *375*, 20190650.
62. Goetz, F.; Jasonowicz, A.; Johnson, R.; Biga, P.; Fischer, G.; Sitar, S. Physiological differences between lean and siscowet lake trout morphotypes: Are these metabolotypes? *Can. J. Fish. Aquat. Sci.* **2014**, *71*, 427–435. [CrossRef]
63. Ballester-Lozano, G.F.; Benedito-Palos, L.; Estensoro, I.; Sitjà-Bobadilla, A.; Kaushik, S.; Pérez-Sánchez, J. Comprehensive biometric, biochemical and histopathological assessment of nutrient deficiencies in gilthead sea bream fed semi-purified diets. *Br. J. Nutr.* **2015**, *114*, 713–726. [CrossRef]
64. Kyneb, A.; Toft, S. Effects of maternal diet quality on offspring performance in the rove beetle tachyporus hypnorum. *Ecol. Entomol.* **2006**, *31*, 322–330. [CrossRef]

Journal of
Marine Science and Engineering

MDPI

Article

Aspects of Reproductive Biology of the European Hake (*Merluccius merluccius*) in the Northern and Central Adriatic Sea (GSA 17-Central Mediterranean Sea)

Michela Candelma [1], Luca Marisaldi [1], Daniela Bertotto [2], Giuseppe Radaelli [2], Giorgia Gioacchini [1], Alberto Santojanni [3], Sabrina Colella [3] and Oliana Carnevali [1,*]

1 Laboratory of Developmental and Reproductive Biology, DiSVA, Università Politecnica delle Marche, 60131 Ancona, Italy; m.candelma@univpm.it (M.C.); l.marisaldi@univpm.it (L.M.); giorgia.gioacchini@univpm.it (G.G.)
2 Department of Comparative Biomedicine and Food Science, University of Padua, 35122 Padova, Italy; daniela.bertotto@unipd.it (D.B.); giuseppe.radaelli@unipd.it (G.R.)
3 National Research Council (CNR), Institute of Biological Resources and Marine Biotechnologies (IRBIM), 60125 Ancona, Italy; alberto.santojanni@cnr.it (A.S.); sabrina.colella@cnr.it (S.C.)
* Correspondence: o.carnevali@staff.univpm.it; Tel.: +39-0712-204-990

Citation: Candelma, M.; Marisaldi, L.; Bertotto, D.; Radaelli, G.; Gioacchini, G.; Santojanni, A.; Colella, S.; Carnevali, O. Aspects of Reproductive Biology of the European Hake (*Merluccius merluccius*) in the Northern and Central Adriatic Sea (GSA 17-Central Mediterranean Sea). *J. Mar. Sci. Eng.* **2021**, *9*, 389. https://doi.org/10.3390/jmse9040389

Academic Editor: Francesco Tiralongo

Received: 26 February 2021
Accepted: 31 March 2021
Published: 7 April 2021

Abstract: The study focused on the macroscopic, histological, and biometric analysis of European hake females in GSA 17 (Central-North Adriatic Sea). From 2013 to 2015, 976 females were collected and analyzed monthly. Though females in spawning conditions were found during the whole year, the trend of GSI showed a peak of the reproductive season from April to July in 2014 and 2015. HSI and K_n reached the highest values in September, after the spawning peaks. In 2013, the trend of these indices did not highlight an evident peak, probably due to an adverse event that occurred in the previous winter in the Adriatic shelf. The length at first maturity (L_{50}) was estimated by macroscopic and histological approaches, resulting in 30.81 cm for the macroscopical length and 33.73 cm for the histological length; both values are higher than the current catching legal size. For the first time in this area, batch and relative fecundity were estimated. Relative fecundity was similar to the Mediterranean and the Atlantic stocks, whereas batch fecundity values were lower compared to other fishing grounds. Overall, the analysis of reproductive parameters plays a fundamental role in the sustainable management of this resource in an area as overfished as the Central-North Adriatic Sea.

Keywords: European hake; *Merluccius merluccius*; fecundity; somatic indices; Adriatic Sea; L_{50}

1. Introduction

In the last several decades, the excessive fishing effort, together with the increase of pollution, poor fishing management, and impairment of marine ecosystems, caused the depletion of fish stocks worldwide. The Food and Agriculture Organization reported that 57.4% of fish stocks are fully exploited, 29.9% are overexploited, and only 12.7% are not fully exploited [1]. In the Mediterranean Sea, the long-lasting, intense fishing pressure applied on fish and invertebrate stocks has led to declining population biomasses [2], which has also been reflected in a reduction of catches for the majority of stocks [3]. Mullon et al. [4], by analyzing the FAO dataset of world fisheries catches for the period 1950–2000, detected that the major collapse occurred for demersal species. Colloca and coworkers [5] assessed the impacts of the fishing pressure in the period 2002–2014 in the Mediterranean Sea and determined that the Central-North Adriatic represents the highest catching area of demersal species in the Mediterranean area.

Among demersal species, the European hake (*Merluccius merluccius*, L. 1758) received attention because it represents one of the principal fishing targets in the Northeast Atlantic Ocean and the Mediterranean Sea. In the Mediterranean Sea, a total landing of 22,547 tons was recorded in 2011 [6,7]; in the Central-North Adriatic Sea, the European hake is one of the leading commercial species [8] and represents 77% of landings from Croatia [9]. To avoid the collapse of hake stock, fishing should be conducted more sustainably by maintaining a spawning stock biomass level that is suitable to guarantee the renewal of the species. In this regard, egg-producing females contribute more to embryonic development success. The acquisition of information on reproductive cycle of the female Adriatic stock, including length at first maturity, fecundity, and spawning cycle, is essential to quantify the reproductive potential of this population [10]. Because of the great commercial importance of the European hake and the critical status of its stocks, several studies were performed in different areas of the Mediterranean Sea related to stock assessment, spawning cycle and fecundity estimation [11–16], feeding habits [17,18], analysis of lipid reserves [19], nursery area [6,20], juvenile recruitment [21], the selectivity of fishing gear [22], fisheries management [2,23,24], and reproductive physiology [25,26]. However, none investigated the fecundity, the spawning cycle, and the proportion of mature individuals at length by a macroscopic and histological approach in the Central-North Adriatic Sea (FAO Geographical Sub-Area 17, according to GFCM division), where this species represents a valuable economic resource.

In this regard, a multiannual survey (2013–2015) was conducted to improve understanding of the reproductive dynamics of European hake in this area. In particular, the work focused on females since the female conditions limit the egg production and the progeny production more than males, and it evaluated the size at first maturity in females using the histological and macroscopic analysis of the gonads. The condition indices such as Le Cren's condition factor (K_n), the hepatosomatic index (HSI), and gonadosomatic index (GSI) were calculated and compared across different ovarian stages, and their changes analyzed in the three years. Furthermore, batch and relative fecundities were estimated together with the analysis of ovarian stages.

This study has never been carried out in this area (GSA 17) and completes the scenario of this important resource of the Italian seas, in order to provide knowledge for a sustainable management.

2. Materials and Methods

2.1. Sampling

A total of 976 females ranging from 13 to 64 cm in total length (TL) were collected aboard commercial bottom trawler fishing vessels in the Northern and Central Adriatic Sea (FAO Geographical Sub-Area 17, according to GFCM division) monthly from January 2013 to December 2015. Guidelines of the Data Collection Framework Regulation (EU Reg.199/2008) followed those established by the Community system for the conservation and sustainable exploitation of fisheries resources under the Common Fisheries Policy (CFP). The sampling procedures did not include any animal experimentation and animal ethics approval was therefore not necessary under the Italian legislation (D.L. 4 March 2014, n. 26, art. 2). For each specimen, the following parameters were recorded: total weight (TW), gutted weight (GW), total length (TL), sex, macroscopic maturity stage of gonads, and gonad and liver weights.

2.2. Reproductive Seasonality and Fish Condition Indices

To evaluate the temporal variation of maturity and condition stage of females, the gonadosomatic index (GSI), hepatosomatic index (HSI), and Le Cren's relative condition factor (K_n) were calculated. These indices were defined by the following equations:

$$\text{GSI}\ (\%) = \text{gonad weight/gutted weight body*100}$$

$$\text{HSI}\ (\%) = \text{liver weight/gutted weight body*10}$$

$$K_n = \text{W}/a\ \text{TL}^b$$

where a and b are the regression parameters of the length–weight relationship, W is gutted weight, and TL is the total length. GSI was evaluated only for the mature females;

313 mature animals were found from developing to regenerating phase. From January to March of 2013, livers were not available for sampling problems. The spawning season was also investigated by analyzing the monthly frequency of ovarian maturity stages. Number of samples of mature females used for GSI, HSI and Kn calculation for each month in three years are reported in Table S1.

2.3. Histological Analysis

173 female specimens at different maturity stages were randomly chosen for histological analysis while choosing a group that was representative for all sizes to confirm the macroscopic classification of ovarian development. Ovarian samples were fixed at 4 °C overnight in formalin at 10% in phosphate-buffered saline (PBS, 0.1 M, pH 7.4), after which they were washed in PBS and stored in 70% ethanol at 4 °C until use. The samples were then dehydrated through a series of graded ethanol, cleared in xylene, and embedded in paraffin wax. Five μm thick consecutive sections were cut using a microtome RM2125 RTS (Leica Biosystems, Wetzlar, Germany) and stained with Meyers's hematoxylin and eosin. Images were acquired using a Zeiss Axio Imager M2 microscope (Zeiss, Oberkochen, Germany). Assessment of the reproductive status of samples was performed according to the method of Brown-Peterson and coworkers [27] but adapted for European hake following the method of Candelma and coworkers [25]. Five ovarian stages were defined: (i) immature; (ii) developing; (iii) spawning capable and actively spawning; (iv) regressing (postspawning); (v) regenerating (spent) (Table 1).

Table 1. Criteria used to determine the maturational status of European hake females.

Ovarian Phases	Macroscopic Morphology	Histological Aspect	Follicle Diameter (μm)
Immature	Orange, semi-transparent	Presence of oogonia (O), primary oocytes (PO). Scarce connective tissue and well compact ovigerous lamellae.	<250
Developing	Small pink but some oocytes visible	Presence of early (vtg1) and middle (vtg2) vitellogenic oocytes. Yolk vesicles form in the ooplasm. Lipid granules occupy a larger cytoplasmic area than yolk droplets. The zona radiata thickens.	250–550
Spawning capable and actively spawning subphase	Large ovaries, oocytes visible macroscopically. An abundance of hydrated oocytes in actively spawning subphase.	Presence of late vitellogenic oocytes (vtg3). Lipid globules occupy a cytoplasmic area like that occupied by yolk granules. Actively spawning subphase: lipid globules fuse into a single larger oil droplet, yolk droplets start to coalescence, the nucleus starts to migrate peripherally to the animal pole (Mn) and POFs. Follicles grow due to water uptake and become transparent (H). Zona radiata is thinner than in previous stages	550–1150
Regressing (postspawning)	Flaccid and small ovaries, blood vessels prominent.	Atresia (any stage) and POFs present. Some CA and/or vtg1, vtg2 oocytes present.	<250
Regenerating	Pinkish and small ovaries that occupy 1/3 of the body cavity, blood vessels reduced but present.	Presence of oogonia (O), primary oocytes (PO), cortical alveoli (CA), and lipid stage oocytes (LS). The ovarian wall is thickened.	<250

The measurement of oocytes was only from oocytes that had been sectioned through the nucleus. The oocyte classification followed Candelma et al. [26] and Murua et al. [28]: oogonia (O); primary oocyte (PO); cortical alveoli (CA); lipid stage (LS); early vitellogenesis (vtg 1); middle vitellogenesis (vtg 2); late vitellogenesis (vtg 3); migrating nucleus (Mn); hydration (H); postovulatory follicles (POFs). Cohen's *k* coefficient [29] was applied to assess the agreement between the histological and macroscopic classification.

2.4. Size at First Maturity (L_{50}) and Fecundity

The proportion of maturity at length (PL) was estimated using both histological and macroscopic data with the following logistic function:

$$PL = \frac{1}{1 + \exp(\alpha + \beta * TL)}$$

where α (intercept) and β (slope) represent the estimated parameters and TL is the total length. All specimens were used for macroscopic investigation; 173 individuals were examined using histological analysis. Females from the developing stage onwards were considered mature. The length at which 50% of the females are mature was computed as $L_{50} = -\alpha/\beta$. The two regression logistic curves of histological and macroscopic data were compared with the likelihood ratio test.

With regards to batch and relative fecundities, 28 females at the actively spawning stage representing the total number of females available were used and the gravimetric method, based on the relationship between ovary weight and oocyte density as described by Murua et al. [30], was applied. Because the distribution of hydrated oocytes does not statistically vary in the ovaries [31], only a lobe of the hydrated ovary was used for analysis. The hydrated oocytes from fresh subsamples were manually counted under a stereomicroscope (Optika, Italy). Batch fecundity was estimated as the average of the hydrated oocyte number integrated over three subsamples multiplied by ovary mass for each specimen with the hydrated oocyte [11]. The relative fecundity was calculated as a ratio of batch fecundity and gutted weight for each fish.

2.5. Statistical Analysis

Statistical differences of the monthly variation of indices for each year and of indices per ovarian stages in each year were determined using a Tukey's multiple test comparison. Both analyses were performed using Prism 6 (GraphPad Software, San Diego, CA, USA). For the determination of size at first maturity, the R statistical environment [32] and the packages "Fisheries Stock Assessment" (FSA) [33], ggplot2 [34], and rel [35] were used. The p-values < 0.05 were considered as significant. Results were expressed as the mean ± SEM.

3. Results

3.1. Reproductive Seasonality and Fish Condition

The percentages of European hake females in different maturity stages observed monthly in the three years indicated that specimens from developing to regressing phase were present throughout the year (Figure 1).

In Figure 2, the GSI trend of mature females during three years of sampling is represented. The GSI values were lowest in 2013 compared to the other two years. In 2013, the highest value was recorded in April with a significant difference only compared to October (Tukey's multiple comparison test; $p < 0.05$) (Figure 2a). Differently from 2013, in 2014 and 2015, the highest peak was observed in June and the lowest in September (Figure 2b,c). In particular, in 2014, the GSI in June was significantly higher than February, September, October, November, and December (Tukey's multiple comparison test; $p < 0.05$) (Figure 2b), and the peak in July was significantly higher than September, December, and February (Tukey's multiple comparison test; $p < 0.05$). In 2015, June was significantly different only compared to September (Tukey's multiple comparison test; $p < 0.05$) (Figure 2c).

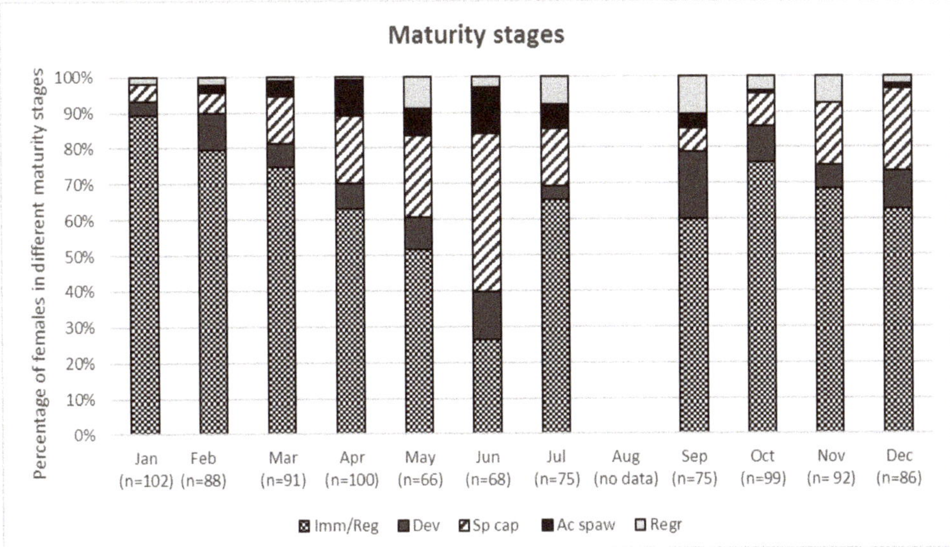

Figure 1. Monthly percentages of European hake female maturity stages. Imm/reg, includes individuals in immature and regenerating stages; dev, developing; spaw, spawning capable; ac spaw, actively spawning, regr, regressing. For their histological characteristic, the immature and regenerating stages were grouped (See Section 3.2).

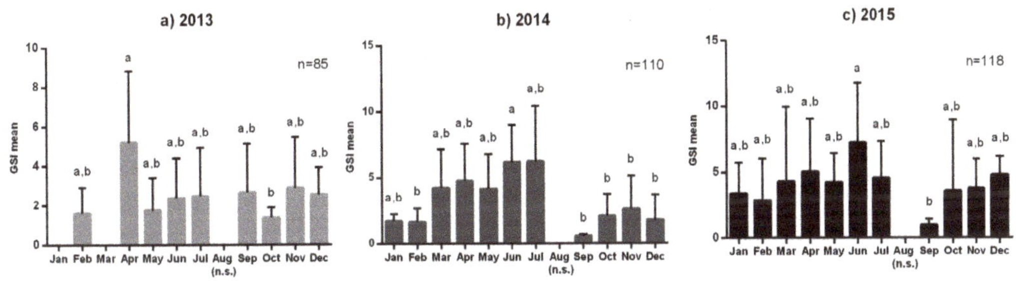

Figure 2. Monthly variation of European hake female gonadosomatic index (GSI) for three years. Statistical differences were described in the result section. In January and March of 2013, no mature females were found. In August of the three years, no samples (n.s.) were sampled. The letters indicate a statistically significant difference among groups ($p < 0.05$) determined by Tukey's multiple comparison test.

The HSI analysis distinguished immature females from mature females (considering from developing to regenerating phase) (Figure 3). The HSI trend varied over the three years. In immature females from 2013, the highest values were in summer months, but significantly different only with September (Tukey's multiple comparison test; $p < 0.05$) (Figure 3a). In 2014, the highest peak was reached in December, significantly different compared to March (Tukey's multiple comparison test; $p < 0.05$) (Figure 3b), whereas in 2015, the highest value was in September and significantly different only compared to March (Tukey's multiple comparison test; $p < 0.05$) (Figure 3c). The lowest value was in March in both 2014 and 2015. In 2013, as in immature females, the mature females were also characterized by an increase of HSI values from May to July (Figure 3d), whereas the lowest measurement was in April significantly diverse compared to July, September, October, November, and December (Tukey's multiple comparison test; $p < 0.05$). Different peaks were recorded during 2014, the highest in December, significantly higher than March–

July (Tukey's multiple comparison test; $p < 0.05$) (Figure 3e). In 2015, the HSI surged in September, significantly different compared to all months (Tukey's multiple comparison test; $p < 0.05$), except to January and December (Figure 3f), and the lowest value was in May.

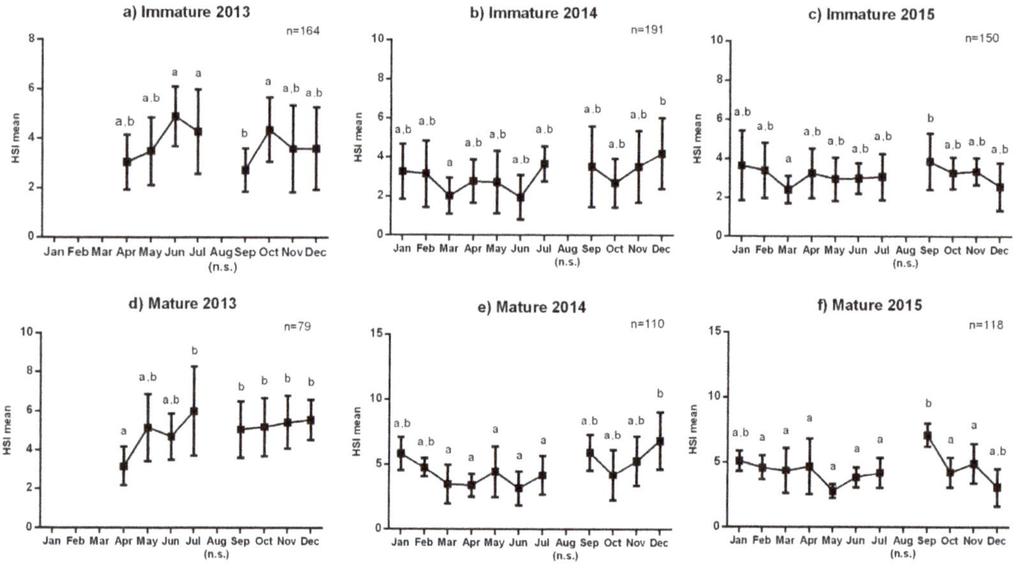

Figure 3. Monthly variation of European hake female hepatosomatic index (HSI) for three years in different ovarian stages. Graphics (**a**–**c**) indicate the females in the immature phase, while graphics (**d**–**f**) show females from developing to regenerating phase. From January to March of 2013, no sampling of the liver was conducted. In August of the three years no samples (n.s.) were sampled. The letters indicate a statistically significant difference among groups ($p < 0.05$) determined by Tukey's multiple comparison test.

The K_n analysis distinguished immature females from mature females (considering from developing to regenerating phase) (Figure 4). In immature individuals collected in 2013, the highest peak was in October and significantly different only compared to April (Tukey's multiple comparison test; $p < 0.05$) (Figure 4a), while in mature animals no significant variation was observed (Figure 4d). The trend of K_n varied mainly in 2014 and 2015, with the highest peak recorded in September for both immature and mature females (Figure 4b–f). In immature individuals captured in 2014, the lowest value was in March and significantly different compared to September and November (Tukey's multiple comparison test; $p < 0.05$) (Figure 4b), whereas mature females showed the lowest value in June with significant differences respect to September and December (Tukey's multiple comparison test; $p < 0.05$) (Figure 4e). In 2015, the K_n of immature females was significantly higher in September compared to the rest of the year except for May, June, July, and October (Tukey's multiple comparison test; $p < 0.05$) (Figure 4c). In mature females captured in 2015, the highest peak of September was significantly different compared to January, April, and May (Tukey's multiple comparison test; $p < 0.05$) (Figure 4f), whereas the value was lowest in December.

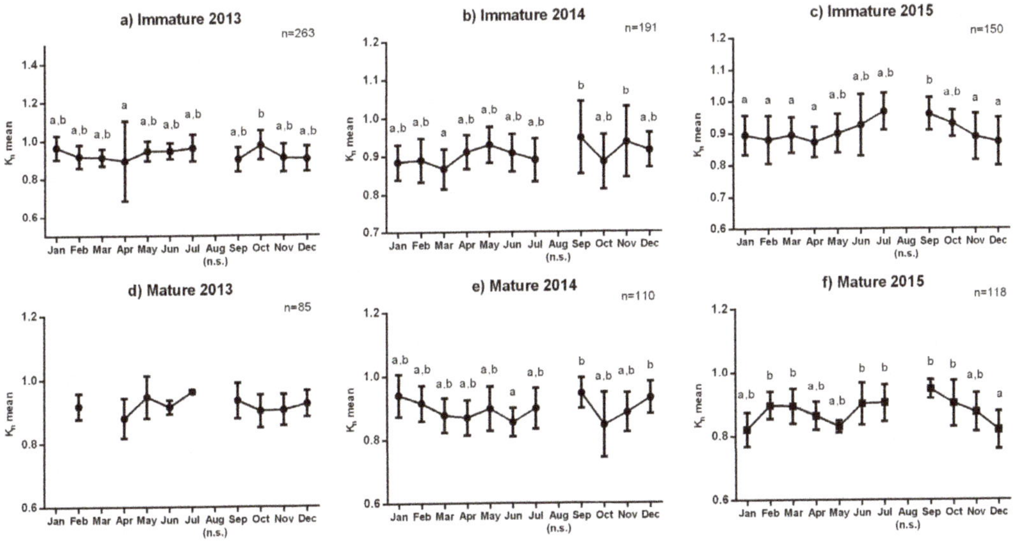

Figure 4. Monthly variation of European hake female Le Cren's condition factor (K_n) for three years in different ovarian stages. Graphics (**a–c**) indicate the females in the immature phase; graphics (**d–f**) show females from developing to regenerating phase. In January and March of 2013, mature individuals were not found. In August of the three years, no samples (n.s.) were sampled. The letters indicate a statistically significant difference among groups ($p < 0.05$) determined by Tukey's multiple comparison test.

The analysis of the condition indices per ovarian stage per three years was macroscopically performed (Figure 5a–c). For their histological characteristic, the immature and regenerating stages were grouped. Regarding the GSI, the significantly highest peak was evidenced in the spawning phase for all three years (Tukey's multiple comparison test; $p < 0.05$) (Figure 5a), the trend increased from imm/reg to the spawning stage for all three years but was significant only in 2013 (Tukey's multiple comparison test; $p < 0.05$). For HSI, the tendency of mean values was lowest in the imm/reg stage for all three years (Tukey's multiple comparison test; $p < 0.05$) (Figure 5b). The other mean values were not significantly different among them within the same year, except for 2014, in which the spawning phase showed a significant decrease (Tukey's multiple comparison test; $p < 0.05$) compared to developing and regressing stages. The variation of K_n was less visible compared to the other two indices in the period analyzed (Figure 5c). The only significant difference (Tukey's multiple comparison test; $p < 0.05$) was evidenced in the spawning capable stage compared to the imm/reg and developing stages in 2014 and only compared to imm/reg phase in 2015.

3.2. Histological Analysis: Ovarian Classification and Patterns

The female reproductive cycle of the European hake consists of five ovarian reproductive stages according to the histological classification (Table 1). Ovaries from immature individuals exhibited compact ovigerous lamellae with groups of oogonia and previtellogenic oocytes (diameter < 250 μm) and an ovarian wall usually thinner than that of the regenerating stage (Figure 6a). Entry into the developing phase is characterized by the appearance of early and the middle vitellogenic oocytes with a diameter ranged 250–550 μm (Figure 6b). The oocytes take up the yolk proteins. The yolk granules multiply and increase in size, forming a densely packed zone in the inner part of the cytoplasm. The oil droplets proliferate, becoming lipid globules and the zona radiata increases in width. The ovaries from spawning specimens were characterized by the presence of fully grown vitellogenic oocytes (vtg3) and/or migrating nucleus (Mn) stage as well as hydrated oocytes (H) (diameter 550–1150 μm) (Figure 6c–e). Nuclear migration and/or hydration distinguish the actively spawning subphase (Figure 6d). The presence of postovulatory follicles in the same ovary with vtg3, OM, and H was observed (Figure 6e). Extensive atresia and a reduced number of vitellogenic oocytes were considered markers of the regressing stage (Figure 6f). The females in the regenerating phase have spawned at least one time in life. Their ovaries are loose and have thickened wall and CA and LS oocytes are present among previtellogenic oocytes (diameter < 250 μm) (Figure 6g).

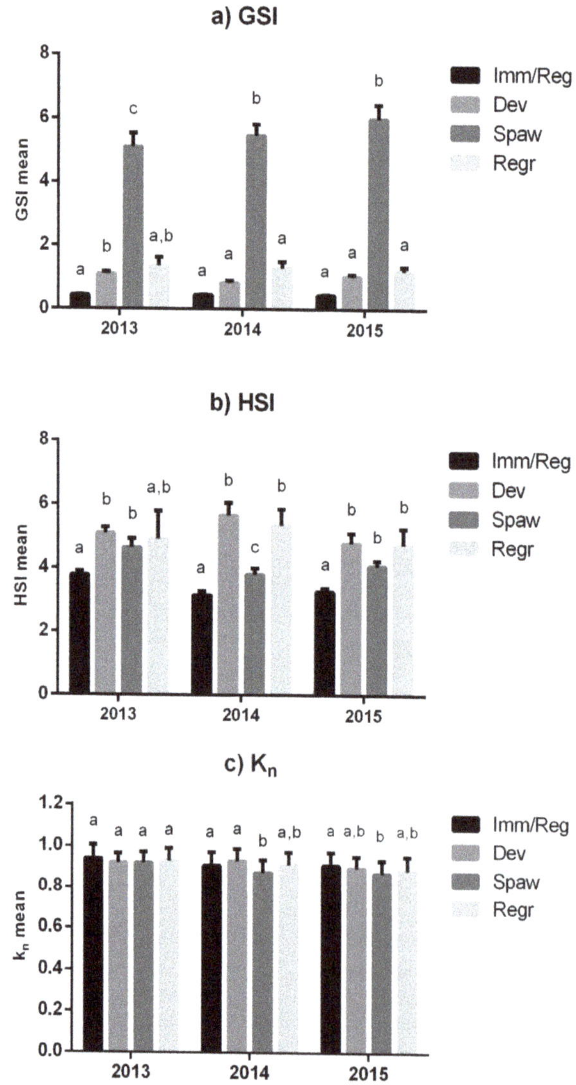

Figure 5. Values of the GSI (**a**), HSI (**b**) and K_n (**c**) for females of European hake per ovarian stages for three years. Imm/reg, includes individuals in immature and regenerating stages; dev, developing; spaw, spawning capable; regr, regressing. The letters indicate a statistically significant difference among groups ($p < 0.05$) determined by Tukey's multiple comparison test.

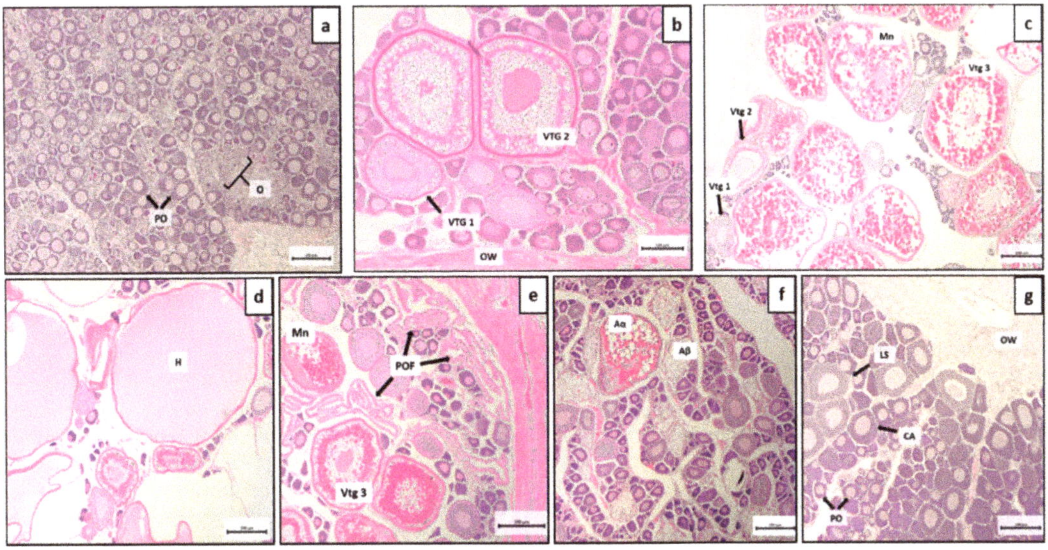

Figure 6. Tissue sections of European hake ovaries at different developmental stages. (**a**) Immature; (**b**) developing; (**c**) spawning capable; (**d**,**e**) actively spawning; (**f**) regressing; (**g**) regenerating. For a,b,g scale bar = 100 µm; for c,d,e,f scale bar = 200 µm. The abbreviations indicate: OW, ovarian wall; O, oogonia; PO, primary oocytes; CA, cortical alveoli; LS, lipid stage; vtg1, vitellogenesis 1; vtg2, vitellogenesis 2; vtg3, vitellogenesis 3; Mn, migrating nucleus; H, hydrated oocytes; POF (postovulatory follicle); Aα, alpha atretic oocyte; Aβ, beta atretic oocyte.

The histological investigation showed a 60.6% similarity with the macroscopic maturity stage classification of the females. Cohen's k was 0.45 (95% confidence interval: 0.35–0.55), which corresponds to a "Moderate" level of agreement [36]. Taking into consideration the macroscopic classification, the percentage of agreement between the histological and macroscopic analysis was different among phases. For the imm/reg phase was 91.8%, for the developing stage was 45.8%, for the spawning phase was 85.3%, and for regressing was 2.86%.

3.3. Size at First Maturity (L_{50}) and Fecundity

L_{50} was estimated macroscopically and histologically to be 30.81 cm and 33.73 cm (Figure 7), respectively.

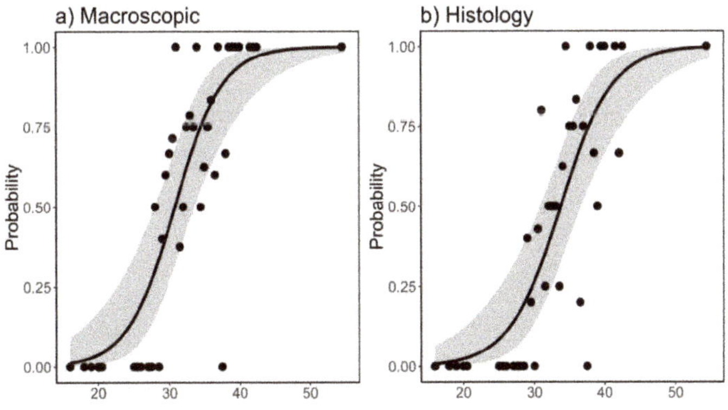

Figure 7. Estimated L_{50} and observed the proportion of mature individuals by 1 cm length classes. (**a**) Macroscopic value; (**b**) histological value. The shaded area indicates a 95% confidence interval.

The estimated parameters of the logistic regression were statistically significant ($p < 0.05$) and summarized in Table 2.

Table 2. Summary of the L50 estimates based on histological and macroscopic data.

	Model Estimates HISTOLOGY			
	Estimate	**Std. Error**	***z* value**	***p*-value**
Intercept	−9.17104	2.00365	−4.577	$p < 0.001$
Total_length_cm	0.27183	0.05925	4.588	$p < 0.0001$
	Model Estimates MACROSCOPIC			
	Estimate	**Std. Error**	***z* value**	***p*-value**
Intercept	−9.15307	2.13655	−4.284	$p < 0.0001$
Total_length_cm	0.29703	0.06505	4.566	$p < 0.0001$

The two curves of L_{50} were not significantly different ($p > 0.05$). The size at which all the females reproduced at least once was 39 cm. The shortest length at which the female specimens entered ovarian development (vitellogenesis) was 26 cm.

Batch fecundity estimation ranged between 7771 (TL 33 cm) and 137,256 (TL 32 cm) hydrated eggs. The shortest total length of a specimen with hydrated eggs in the ovary was 29 cm (batch fecundity of 45,004 eggs) and the longest was 42 cm (batch fecundity 70,519 eggs). The mean value of a relative fecundity was 205 eggs/g of body weight, the lowest value of relative fecundity was 34 eggs/g of body weight found in a female with TL of 33 cm, and the highest value of relative fecundity was 573 eggs/g of body weight derived from a female with TL of 32 cm.

4. Discussion

One of the targets of fisheries management is to conserve enough reproductive potential in a population to allow for sustainable exploitation [37]. To achieve this, the reproductive parameters, such as length at first maturity, fecundity, and spawning cycle determination, are usually applied as health indicators of the stock. The present study is the first attempt to highlight an exhaustive knowledge of the reproductive biology of *Merluccius merluccius* in the Northern and Central Adriatic Sea (GSA 17) to complete the scenario of knowledge in the Italian seas.

The easiest, cheapest, and most direct approach to investigate some aspects of reproductive biology is a macroscopic evaluation of the maturity stages of the gonads, but it is not always an accurate method [16,38,39]. This study considered the somatic indices to evaluate the reproductive state of the European hake. We focused on different indices, namely the gonadosomatic index (GSI), the hepatosomatic index (HSI), and Le Cren's condition factor (K_n) along three consecutive years. The GSI trend calculated on mature specimens sampled from the GSA17 indicated that the hake reproductive season reached a peak from March to July, even though a presence of spawning females was observed throughout the year, confirming that a protracted spawning period is typical for this species [11,40,41]. Unlike 2014 and 2015, in 2013, a lack of the GSI peak during the reproductive season was observed. The possible reason for this could be related to an exceptional event of dense water that occurred in the winter of 2012 in the Adriatic shelf [42]. Such a phenomenon was characterized by high-velocity currents that caused a cascading event leading to the transport of more than 50% of water volumes from the Northern Basin to the Southern area of the Adriatic Sea [42–44]. Directly or indirectly, unfavorable conditions for the successfulness of the hake reproduction could have emerged, as previously suggested by Recasens et al. [11], in which dense water formation caused intense cascading events shifting the hake reproduction in the Catalan Sea by several months. Thus, the oceanographic events that took place in the Adriatic Sea in 2012 could have negatively impacted the stock of the European hake in 2013 in the area. Regarding the GSI value in different ovarian stages, the highest mean value was reached at the spawning stage, confirming its correlation with the sexual maturity of gonads.

In the European hake, lipids are stored mainly in the liver, confirming the critical role of this organ for energy storage in this species [19]. The analysis of HSI represents a good proxy of the energy reserve available in fish. In the Central-North Adriatic Sea, HSI showed different trends within the three years studied in both immature and mature females. In 2014 and 2015 for all ovarian phases, the highest recorded values of HSI after summer are probably due to the increase in the food availability for this demersal predator during this period. The energy stored during richer months foodwise is not followed by reproductive activity, allowing the European hake to survive during the winter period, as previously reported in Galician shelf stock [45]. In 2013, the HSI reached the greatest value of the year in the summer months, probably because the absence of the reproductive activity allowed an incessant accumulation of lipids. Based on ovarian developmental stages, the lowest level of HSI was obtained in the imm/reg stage, demonstrating that juvenile fish use lipid energy for growth and do not accumulate them in the liver [46].

Together with HSI, Le Cren's condition factor indicates the energy storage in fishery species. It is given by weight and length data and assumes that heavier fish of a given length are in a better condition [47]. In 2014 and 2015, the K_n showed the highest values in September both for immature and mature individuals, just like HSI. In 2013, the autumn peak shifted one month only for immature females, whereas no variation was shown in mature animals, so we speculated that like for HSI, the K_n also indicates an incessant accumulation of energy in fish during such a period. In the spawning capable phase, the low levels of K_n occurred in 2014 and 2015; only HSI in 2014 could confirm that European hake uses part of the energy accumulated in previous months for reproduction if it is necessary.

The assessment of maturity stages of the ovaries is a crucial step for the estimation of the L_{50}, and errors might lead to subsequent biased estimation of the spawning stock biomass [38,39,48]. In this contest, an accurate assessment of sexual maturity is an essential component of effective fisheries management. For European hake, some sexual maturity classifications exist, but often in these studies, a different terminology was used for the oocyte stages or ovarian phases [12,16,28,49]. Furthermore, some authors proposed only an oocyte classification, whereas other authors suggested a macroscopic and histological scale, but without a proper calibration between visual (macroscopic) and histological (microscopic) staging. We used the oocyte stage classification proposed by Candelma et al. [26] and Murua et al. [28] and the standardized terminology for ovarian phases described by Brown-Peterson et al. [27] to obtain a clear and accurate scale on ovarian and oocyte stages validated by the similarity between macroscopic and histological approach. The "Moderate" level of agreement in the assignment of reproductive stages between histological and macroscopic investigation revealed that our macroscopic classification for European hake does not suffer from relatively high error rates. An exception was the regressing phase, whose percentage of agreement was weak. Overall, our scale appeared an accurate tool for the sexual maturity determination by the macroscopic method, notwithstanding some stages require (i.e., the regressing phase and the immature stages) histological analysis to avoid misclassification. Moreover, the histology is the most suitable method for the studies on reproductive biology and it is strictly recommended a regular calibration (over the years) between the two approaches. Histological analysis confirmed that *M. merluccius* is a batch spawner with an asynchronous type of ovarian development.

The L_{50} is a fundamental parameter of fish population dynamics studies and plays a crucial role in species management purposes. In the present study, the length at first maturity was estimated as 30.81 cm by macroscopic evaluation of gonads maturity and 33.73 cm by histological analysis. The L_{50}, determined in this work, is comparable to the value of 30.5 cm obtained in a previous study of Zupanovic and Jardas [50] in the same area, while Alegria Hernandez and Jukic [51] estimated an L_{50} equals to 31.3 cm in total length, but both only by macroscopic inspection. In the Southern Adriatic, the size of the first maturity calculated using the macroscopic method was 31.95 cm [16]. In comparison with previous studies on length at first maturity in the Mediterranean Sea, the values estimated were similar to the measure reported in the present study. On the Egyptian coast, Al-Absawy [14] found an L_{50} of 32.5 cm, while in the Algerian coast [13], it was 30.6 cm. The first maturity size is generally affected by variations of several environmental factors that characterize the different areas, such as the abundance and distribution of local populations, competition for space, and availability of nourishment. Furthermore, the overexploitation of this resource could explain the differences in this parameter that reduces the spawning biomass in different areas [12,16]. The L_{50} of 33.73 cm resulting from histological analysis evidenced that the macroscopic approach underestimates this value, but the two curves were not significantly different, indicating that the histological examination is not strictly necessary.

The L_{50} found in the present study contrasts the decision regulated by Annex III of the Council Regulation (EC) N° 1967/2006, in which the minimum legally allowed fishing size is 20 cm, implying the catch of sexually immature individuals. This excessive fishing effort could lead to a likely decrease in the stock in the area in upcoming years. Accordingly, as reported by Anderson et al. [52], the reduction of the age of a stock and average body size can reduce the ability of exploited species to survive to annual environmental variation.

In the present study, for the first time, batch and relative fecundities in the Northern-Central Adriatic Sea were analyzed. The mean value of a relative fecundity reported in this study (205 eggs/g of body weight) seemed comparable with the previous estimates reported in Catalan waters (204.29 eggs/g of body weight), the North Tyrrhenian (202.35 eggs/g of body weight) [11], and in the Eastern-Central Atlantic (228.33 eggs/g of body weight) [12], yet it was lower than the value estimated in the Central Tyrrhenian and Southern Adriatic Sea (281.6 eggs/g of body weight GSA10 and 262.2 eggs/g of body weight GSA18) [16]. The batch fecundity estimation ranged between 7771 and 137,256 eggs. It was lower compared to the Southern Adriatic Sea and Central Tyrrhenian, probably due to the major dimensions of reproductive animals in these areas compared to Northern and Central Adriatic Sea [16]. In fact, as reported by Recasens et al. [11], El Habouz et al. [12], Carbonara et al. [16] and Korta et al. [53], the relationship between fecundity estimations and hake size influences the total of eggs number per female that increases with size, weight, and gonad weight.

In conclusion, the scenario depicted in the present study evidenced that smaller mature females released fewer eggs per spawn than bigger mature females. As reported by Working Group on Stock Assessment of Demersal Species (WGSAD) [54], the European hake stock of Adriatic Sea results to be overfished and the overexploitation together with a low number of spawned eggs, and unfavorable environmental events could lead to a consequent decrease of the fish resource with the collapse of the stock. To avoid the collapse of European hake, especially in the Central-North Adriatic Sea, continuous monitoring programs, estimation of reproductive potential, and the stock assessment are essential. The easiest, cheapest, and most direct tool to investigate the reproductive stage of European hake, necessary for the determination of reproductive potential, could be represented by the maturity scale provided in this study. Furthermore, the reproductive cycle, the analysis of somatic indices, the L_{50} estimations, and the estimation of fecundity represent valuable information for a scientific decision-making process to establish suitable management measures aimed at tackling the continuing stock decline of European hake in an overfished area as Central-North Adriatic Sea. Finally, the result highlighted in the present study suggested increasing the minimum legal catch size as a tool to preserve this resource in this area.

Supplementary Materials: The following are available online at https://www.mdpi.com/article/10.3390/jmse9040389/s1.

Author Contributions: M.C. contributed to the acquisition of data, analysis of data, and drafting of the manuscript. S.C. was involved in contributions to conception and design of the manuscript, acquisition of data, and analysis of data and in revising it critically for important intellectual content. L.M. contributed to the analysis and interpretation of data. A.S. contributed to the conception and design of the project and to revising it critically for important intellectual content. D.B., G.R., and G.G. contributed to the analysis of data and to revising it critically for important intellectual content. O.C. conceived the experimental design, contributed to the analysis of data, and revised it critically for important intellectual content. All authors gave the final approval of the version to be published. Each author has participated in the work to take public responsibility for appropriate portions of the content and agreed to be accountable for all aspects of the work in ensuring that questions related to the accuracy or integrity of any part of the work are appropriately investigated and resolved. All authors have read and agreed to the published version of the manuscript.

Funding: This work was partially supported by the Italian Ministry of Agricultural, Food, and Forestry Policies (MiPAAF) and European Union (Italian National Programs 2011–2013 and 2014–2016, in the ambit of Data Collection Framework).

Institutional Review Board Statement: The sampling procedures did not include any animal experimentation and animal ethics approval was therefore not necessary, under the Italian legislation (D.L. 4 March 2014, n. 26, art. 2).

Informed Consent Statement: Not applicable.

Data Availability Statement: Data available on request due to restrictions, e.g., privacy or ethical. The data presented in this study are available on request from the corresponding author.

Acknowledgments: The authors wish to thank the Filippo Domenichetti and Camilla Croci of Nation-al Research Council (CNR) Institute of Biological Resources and Marine Biotechnologies (IRBIM) Ancona, Italy, and Captain Giordano and crew of the "Orizzonte" vessel for their support in sampling.

Conflicts of Interest: Sabrina Colella, Alberto Santojanni, and Oliana Carnevali contributed equally to this work. The authors declare that there is no conflict of interest.

References

1. FAO. Review of the state of world marine fishery resources. *FAO Fish. Aquac. Tech. Pap.* **2011**, *569*, 334.
2. Colloca, F.; Cardinale, M.; Maynou, F.; Giannoulaki, M.; Scarcella, G.; Jenko, K.; Bellido, J.M.; Fiorentino, F. Rebuilding Mediterranean fisheries: A new paradigm for ecological sustainability. *Fish Fish.* **2013**, *14*, 89–109. [CrossRef]
3. Tsikliras, A.C.; Dinouli, A.; Tsiros, V.-Z.; Tsalkou, E. The Mediterranean and Black Sea Fisheries at Risk from Overexploitation. *PLoS ONE* **2015**, *10*, e0121188. [CrossRef]
4. Mullon, C.; Freon, P.; Cury, P. The dynamics of collapse in world fisheries. *Fish Fish.* **2005**, *6*, 111–120. [CrossRef]
5. Colloca, F.; Scarcella, G.; Libralato, S. Recent trends and impacts of fisheries exploitation on Mediterranean stocks and ecosystems. *Front. Mar. Sci.* **2017**, *4*, 244. [CrossRef]
6. Druon, J.N.; Fiorentino, F.; Murenu, M.; Knittweis, L.; Colloca, F.; Osio, C.; Mérigot, B.H.; Garofalo, G.; Mannini, A.; Jadaud, A.H.; et al. Modelling of European hake nurseries in the Mediterranean Sea: An ecological niche approach. *Prog. Oceanogr.* **2015**, *130*, 188–204. [CrossRef]
7. Cerviño, S.; Domínguez-Petit, R.; Jardim, E.; Mehault, S.; Piñeiro, C.; Saborido-Rey, F. Impact of egg production and stock structure on MSY reference points and its management implications for southern hake (*Merluccius merluccius*). *Fish. Res.* **2013**, *138*, 168–178. [CrossRef]
8. Sabatella, E.C.; Colloca, F.; Coppola, G.; Fiorentino, F.; Gambino, M.; Malvarosa, L.; Sabatella, R. Key Economic Characteristics of Italian Trawl Fisheries and Management Challenges. *Front. Mar. Sci.* **2017**, *4*, 1–10. [CrossRef]
9. Grati, F.; Aladžuz, A.; Azzurro, E.; Bolognini, L.; Car, P.; Spedicato, M.T.; Stagličić, N.; Vrgoć, N.; Zerem, N.; Arneri, E.; et al. Seasonal dynamics of small-scale fisheries in the Adriatic Sea. *Mediterr. Mar. Sci.* **2018**, *19*, 21–35. [CrossRef]
10. Murua, H.; Saborido-Rey, F. Female reproductive strategies of marine fish species of the North Atlantic. *J. Northwest Atl. Fish. Sci.* **2003**, *33*, 23–31. [CrossRef]
11. Recasens, L.; Chiericoni, V.; Belcari, P. Spawning pattern and batch fecundity of the European hake (*Merluccius merluccius* (Linnaeus, 1758)) in the western Mediterranean. *Sci. Mar.* **2008**, *72*, 721–732. [CrossRef]
12. El Habouz, H.; Recasens, L.; Kifani, S.; Moukrim, A.; Bouhaimi, A.; El Ayoubi, S. Maturity and batch fecundity of the European hake (*Merluccius merluccius*, Linnaeus, 1758) in the eastern central Atlantic. *Sci. Mar.* **2011**, *75*, 447–454. [CrossRef]
13. Bouaziz, A.; Bennoui, A.; Djabali, F.; Maurin, C. Reproduction du merlu *Merluccius merluccius* (Linnaeus, 1758) dans la région de Bou-Ismail. *CIHEAM Cah. Opt. Méditerr.* **1998**, *35*, 109–117.

14. Al-Absawy, M.A. The reproductive biology and the histological and ultrastructural characteristics in ovaries of the female gadidae fish *Merluccius merluccius* from the Egyptian Mediterranean water. *Afr. J. Biotechnol.* **2010**, *9*, 2544–2559.
15. Nannini, N.; Pinna, D.; Chiericoni, V.; Biagi, F.; Belcari, P. Ciclo ovarico di *Merluccius merluccius* (Linnaeus, 1758) nel Mar Tirreno settentrionale. *Biol. Mar. Mediterr.* **2001**, *8*, 745–748.
16. Carbonara, P.; Porcu, C.; Donnaloia, M.; Pesci, P.; Sion, L.; Spedicato, M.T.; Zupa, W.; Vitale, F.; Follesa, M.C. The spawning strategy of European hake (*Merluccius merluccius*, L. 1758) across the Western and Central Mediterranean Sea. *Fish. Res.* **2019**, *219*, 105333. [CrossRef]
17. Carpentieri, P.; Colloca, F.; Cardinale, M.; Belluscio, A.; Ardizzone, G.D. Feeding habits of European hake (*Merluccius merluccius*) in the central Mediterranean Sea. *Fish. Bull.* **2005**, *103*, 411–416.
18. Carrozzi, V.; Di Lorenzo, M.; Massi, D.; Titone, A.; Ardizzone, G.; Colloca, F. Prey preferences and ontogenetic diet shift of European hake *Merluccius merluccius* (Linnaeus, 1758) in the central Mediterranean Sea. *Reg. Stud. Mar. Sci.* **2019**, *25*, 100440. [CrossRef]
19. Lloret, J.; Demestre, M.; Sanchez-Pardo, J. Lipid (energy) reserves of European hake (*Merluccius merluccius*) in the north-western Mediterranean. *Vie Milieu-Life Environ.* **2008**, *58*, 75–85.
20. Cantafaro, A.; Ardizzone, G.; Enea, M.; Ligas, A.; Colloca, F. Assessing the importance of nursery areas of European hake (*Merluccius merluccius*) using a body condition index. *Ecol. Indic.* **2017**, *81*, 383–389. [CrossRef]
21. Ligas, A.; Colloca, F.; Lundy, M.G.; Mannini, A.; Sartor, P.; Sbrana, M.; Voliani, A.; Belcari, P. Modeling the growth of recruits of European hake (*Merluccius merluccius*) in the northwestern Mediterranean Sea with generalized additive models. *Fish. Bull.* **2015**, *113*, 69–81. [CrossRef]
22. Deniz, T.; Göktürk, D.; Ateş, C. Selectivity parameters of European hake gillnets for target and by-catch species with a perspective on small-scale fisheries management in the Sea of Marmara, Turkey. *Reg. Stud. Mar. Sci.* **2020**, *33*, 100934. [CrossRef]
23. Angelini, S.; Hillary, R.; Morello, E.B.; Plagányi, É.E.; Martinelli, M.; Manfredi, C.; Isajlović, I.; Santojanni, A. An Ecosystem Model of Intermediate Complexity to test management options for fisheries: A case study. *Ecol. Model.* **2016**, *319*, 218–232. [CrossRef]
24. Tsagarakis, K.; Palialexis, A.; Vassilopoulou, V. Mediterranean fishery discards: Review of the existing knowledge. *ICES J. Mar. Sci.* **2014**, *71*, 1219–1234. [CrossRef]
25. Candelma, M.; Fontaine, R.; Colella, S.; Santojanni, A.; Weltzien, F.-A.; Carnevali, O. Gonadotropin characterization, localization and expression in the European hake (*Merluccius merluccius*). *Reproduction* **2017**, *153*, 123–132. [CrossRef] [PubMed]
26. Candelma, M.; Valle, L.D.; Colella, S.; Santojanni, A.; Carnevali, O. Cloning, characterization, and molecular expression of gonadotropin receptors in European hake (*Merluccius merluccius*), a multiple-spawning species. *Fish Physiol. Biochem.* **2018**, *44*, 895–910. [CrossRef] [PubMed]
27. Brown-Peterson, N.J.; Wyanski, D.M.; Saborido-Rey, F.; Macewicz, B.J.; Lowerre-Barbieri, S.K. A standardized terminology for describing reproductive development in fishes. *Mar. Coast. Fish.* **2011**, *3*, 52–70. [CrossRef]
28. Murua, H.; Motos, L.; Lucio, P. Reproductive modality and batch fecundity of the European hake (*Merluccius merluccius* L.) in the Bay of Biscay. *Calif. Coop. Ocean. Fish. Investig. Rep.* **1998**, *39*, 196–203.
29. Cohen, J. A Coefficient of Agreement for Nominal Scales. *Educ. Psychol. Meas.* **1960**, *20*, 37–46. [CrossRef]
30. Murua, H.; Kraus, G.; Saborido-Rey, F.; Witthames, P.R.; Thorsen, A.; Junquera, S. Procedures to estimate fecundity of marine fish species in relation to their reproductive strategy. *J. Northwest Atl. Fish. Sci.* **2003**, *33*, 33–54. [CrossRef]
31. Murua, H.; Lucio, P.; Santurtún, M.; Motos, L. Seasonal variation in egg production and batch fecundity of European hake *Merluccius merluccius* (L.) in the Bay of Biscay. *J. Fish Biol.* **2006**, *69*, 1304–1316. [CrossRef]
32. R Core Team. R: A Language and Environment for Statistical Computing. R Foundation for Statistical Computing. Available online: https://www.r-project.org/ (accessed on 1 April 2021).
33. Ogle, D.; Wheeler, P.; Dinno, A. FSA: Fisheries stock analysis. In *R Package Version 0.8.25*; Available online: https://github.com/droglenc/FSA (accessed on 1 April 2021).
34. Wickham, H. *ggplot2: Elegant Graphics for Data Analysis*; Springer: New York, NY, USA, 2016.
35. Lo Martire, R. R package version 1.3.1. *Reliab. Coeff.* **2017**. Available online: https://CRAN.R-project.org/package=rel (accessed on 1 April 2021).
36. Landis, J.R.; Koch, G.G. The Measurement of Observer Agreement for Categorical Data. *Biometrics* **1977**, *33*, 159–174. [CrossRef] [PubMed]
37. Domínguez-Petit, R.; Korta, M.; Saborido-Rey, F.; Murua, H.; Sainza, M.; Piñeiro, C. Changes in size at maturity of European hake Atlantic populations in relation with stock structure and environmental regimes. *J. Mar. Syst.* **2008**, *71*, 260–278. [CrossRef]
38. Vitale, F.; Svedäng, H.; Cardinale, M. Histological analysis invalidates macroscopically determined maturity ogives of the Kattegat cod (*Gadus morhua*) and suggests new proxies for estimating maturity status of individual fish. *ICES J. Mar. Sci.* **2006**, *63*, 485–492. [CrossRef]
39. Marisaldi, L.; Basili, D.; Candelma, M.; Sesani, V.; Pignalosa, P.; Gioacchini, G.; Carnevali, O. Maturity assignment based on histology-validated macroscopic criteria: Tackling the stock decline of the Mediterranean swordfish (*Xiphias gladius*). *Aquat. Conserv. Mar. Freshw. Ecosyst.* **2019**, *29*, 1–12. [CrossRef]
40. Murua, H.; Motos, L. Reproductive strategy and spawning activity of the European hake *Merluccius merluccius* (L.) in the Bay of Biscay. *J. Fish Biol.* **2006**, *69*, 1288–1303. [CrossRef]

41. Recasens, L.; Lombarte, A.; Morales-Nin, B.; Tores, G.J. Spatiotemporal variation in the population structure of the European hake in the NW Mediterranean. *J. Fish Biol.* **1998**, *53*, 387–401. [CrossRef]
42. Mihanović, H.; Vilibić, I.; Carniel, S.; Tudor, M.; Russo, A.; Bergamasco, A.; Bubić, N.; Ljubešić, Z.; Viličić, D.; Boldrin, A.; et al. Exceptional dense water formation on the Adriatic shelf in the winter of 2012. *Ocean Sci.* **2013**, *9*, 561–572. [CrossRef]
43. Janekovic, I.; Mihanovic, H.; Vilibic, I.; Tudor, M. Extreme cooling and dense water formation estimates in open and coastal regions of the Adriatic Sea during the winter of 2012. *J. Geophys. Res. Ocean.* **2014**, *119*, 3200–3218. [CrossRef]
44. Benetazzo, A.; Bergamasco, A.; Bonaldo, D.; Falcieri, F.M.; Sclavo, M.; Langone, L.; Carniel, S. Response of the Adriatic Sea to an intense cold air outbreak: Dense water dynamics and wave-induced transport. *Prog. Oceanogr.* **2014**, *128*, 115–138. [CrossRef]
45. Dominguez-Petit, R.; Saborido-Rey, F.; Medina, I. Changes of proximate composition, energy storage and condition of European hake (*Merluccius merluccius*, L. 1758) through the spawning season. *Fish. Res.* **2010**, *104*, 73–82. [CrossRef]
46. Lloret, J.; Shulman, G.; Love, M.R. *Condition and Health Indicators of Exploited Marine Fishes*; Wiley Blackwell: Hoboken, NJ, USA, 2014; ISBN 978-0-470-67024-8.
47. Jones, R.E.; Petrell, R.J.; Pauly, D. Using modified length-weight relationships to assess the condition of fish. *Aquac. Eng.* **1999**, *20*, 261–276. [CrossRef]
48. Flores, A.; Wiff, R.; Ganias, K.; Marshall, C.T. Accuracy of gonadosomatic index in maturity classification and estimation of maturity ogive. *Fish. Res.* **2019**, *210*, 50–62. [CrossRef]
49. Lucio, P.; Murua, H.; Santurtun, M. Growth and reproduction of hake (*Merluccius merluccius*) in the Bay of Biscay during the period 1996–1997. *Ozeanografika* **2000**, *3*, 325–354.
50. Zupanovic, S.; Jardas, I. A contribution to the study of biology and population dynamics of the Adriatic hake, *Merluccius merluccius* (L.). *Acta Adriat.* **1986**, *27*, 97–146.
51. Alegria Hernandez, V.; Jukic, S. Abundance dynamics of the hake (*Merluccius merluccius* L.) from the middle Adriatic sea. *Bull. l'Inst. Océanogr.* **1992**, *11*, 151–161.
52. Anderson, C.N.K.; Hsieh, C.; Sandin, S.A.; Hewitt, R.; Hollowed, A.; Beddington, J.; May, R.M.; Sugihara, G. Why fishing magnifies fluctuations in fish abundance. *Nature* **2008**, *452*, 835–839. [CrossRef]
53. Korta, M.; Dominguez-Petit, R.; Murua, H.; Saborido-Rey, F. Regional variability in reproductive traits of European hake Merluccius merluccius L. populations. *Fish. Res.* **2010**, *104*, 64–72. [CrossRef]
54. FAO. Working Group on Stock Assessment of Demersal Species (WGSAD). *Gen. Fish. Comm. Mediterr.* **2017**. Available online: http://www.fao.org/gfcm/reports/technical-meetings/detail/en/c/1105308/ (accessed on 1 April 2021).

this rapid progress has greatly and negatively increased the impact on the environment and on human health.

An economy based on encouraging the consumption of goods and services promotes a progressive increase in anthropogenic pressure on the environment, due to the increasing loads of substances of all kinds released into the air, water and soil. Exposure to these contaminants can cause harmful effects on the health of organisms and humans.

One of the most dramatic aspects that humans will have to face himself in the immediate future is represented by waters pollution, determined by the growing release of heavy metals which, being unable to be degraded or destroyed, tend to accumulate in the environment and in organisms with consequent and substantial risks both for the various living forms and for humans who consume foods with a content of significantly high metals [1].

The effects of heavy metal pollution on the environment, on the conservation of biodiversity and on human health are now well known [1]. Immediate actions are therefore necessary both aimed at preventing the continuation of uncontrolled discharges and at imposing compliance with current regulations in terms of protection of the environment and public health. These interventions should be preceded by an environmental biomonitoring program that allows rapid identification of the risk areas.

In biomonitoring programs, the choice of species to be used is particularly important. In recent decades among the new generation bioindicators, fish have been taking on a growing interest in assessing environmental quality in various aquatic ecosystems, for their ability of bioaccumulation and biomagnification along the trophic chain [2,3]. It is also necessary to remember that fish are a fundamental component in human nutrition, representing a significant source of protein, polyunsaturated fatty acids and micronutrients. However, through the consumption of fish products, humans are thus exposed to various contaminants.

Heavy metals are elements with a characteristic shiny appearance and a good electric conductivity. In chemical reactions, heavy metals frequently occur as cations [4]. However, non-heavy elements, such as aluminum (density 1.5), can be dangerous under particular conditions. Indeed, it becomes a powerful toxic if placed in acidic waters [4].

From an environmental point of view, the term heavy metals are commonly referred to all metals or non-metals which represent a danger to health and to the environment. They are substances present in the environment as constituent elements of the same, as well as introduced following industrial emissions. They can undergo biological and chemical transformations, which entail their accumulation in the environment and in organisms, both vegetable and animal, thus manifesting their deleterious action. Although currently exposure to anthropogenic sources is of prevailing toxicological importance, exposure to natural sources has proven to be fundamental for the development, in living organisms, of detoxification mechanisms, elimination and use aimed at reducing the danger of metals. These mechanisms allow some animal species to withstand high metal concentrations, which vice versa can be toxic to others, without suffering any damage [5]. Some metals are required by organisms in limited quantities; in particular, Zn, Cu, Fe, and Mg, even if present at low concentrations, perform a series of fundamental activities for the cell behaving as essential micronutrients and participating in numerous biochemical processes responsible for cell growth and life [6].

Zinc, for example, is involved in the replication, transcription and translation processes [7,8], acts as a cofactor for over 200 metalloenzymes [9] and performs regulatory functions, as in the case of modulation of synaptic transmission [10]. Copper, in low concentrations, is essential for breathing, for defense against free radicals and for the synthesis and release of neurotransmitters [11]. Other metals, such as Cd, Cr, Al, Hg and Pb are not normally present in the cells not even in traces and therefore, even at low concentrations, they can be very toxic. These metals can cause delays in embryonic development [12,13], in growth [14–17], as well as a long series of pathologies, including cancer [18,19]. Fish species, for example, hardly eliminate the absorbed mercury and the metal halving times vary from 6 months for mussels to 2 years for pike. The accumulation in fish is greater in muscle tissue than in adipose tissue and about 90–99% of the mercury present in fish is in the form of methylmercury, an extremely toxic form of this metal [20]. Therefore, fish in risk assessment studies are useful bioindicators and can represent an early warning system of environmental damage, which can also be used for the assessment of potential risks to human health.

Merluccius merluccius (Linnaeus, 1758) is a demersal fish whose depth range normally extends from 70 to 400 m, although it can be found in shallower waters and up to about 1000 m depth. Its distribution range extends from the eastern Atlantic (from Norway and Iceland to Mauritania) to the Mediterranean Sea and southern part of the Black Sea. Unlike the small specimens (<14 cm TL), which feed mainly on euphausiids and mysids, the larger specimens (>32 cm TL) of the European Hake, *M. merluccius*, are ichthyophagous [21]. Furthermore, this species has been successfully used as bioindicator [22]. Hence, considering the commercial importance of this predator, its position in the trophic web and its ability to be used as a good bioindicator of marine water pollution, our choice to use *M. merluccius* in the current study.

In this perspective, our work aims to evaluate the impact of metals/metalloids contamination in the commercial fish species *M. merluccius*, commonly known as "European Hake", sampled in the Ionian Sea. Previous research in this area of study report a total metal load higher in pelagic fish than demersal and benthic ones [23,24]. Although fish species resulted stressed by environmental conditions with a certain degree of oxidative stress in liver tissue [23], from a chemical point of view, the analyzed species were healthy for human consumption and the human risk to develop chronic systemic and carcinogenic effects due to their consumption was low [24]. Nevertheless, studies focused on the biomonitoring of the Sicilian ionic coast highlighted a significant metal load in areas subject to heavy anthropogenic pressure, particularly the Augusta coastal water, where it is hosted the largest industrial complex of Sicily [25,26].

The present study has been based on analysis of the gastrointestinal content to evaluate the presence of metals/metalloids; quantitative and qualitative analysis of metals/metalloids in the tissues; analysis of the toxicological effects of metals/metalloids through biomarkers detection (Heat Shock Proteins 70 and Metallothioneis 1) and histological analysis.

2. Materials and Methods

For this study, 20 specimens (10 males and 10 females) of *M. merluccius* fished in Food and Agriculture Organization of the United Nations (FAO) area 37 (Marzamemi, southeastern Sicily, Ionian Sea) were analyzed. Fish size ranged from 35 to 45 cm total length (TL), and were captured with longlines by local fishermen and transported fresh to the laboratory for analyses.

2.1. Metals and Metalloids Analysis

The method for extraction and quantification of metals and metalloids is described by Copat et al. [27]. Briefly, aliquots of 0.5 g of muscle, liver, gills and gastrointestinal content of fish were acid digested with 6 mL of 65% nitric acid (HNO_3) (Carlo Erba) and 2 mL of 30% peroxide hydrogen (H_2O_2-Carlo Erba) in a microwave system (Ethos Touch Control, Milestone S.r.l., Italy). Analytical determination of arsenic (As), cadmium (Cd), cobalt (Co), chromium (Cr), copper (Cu), lead (Pb), mercury (Hg), manganese (Mn), nickel (Ni), vanadium (V), selenium (Se), antimony (Sb) and zinc (Zn) was performed with an ICP-MS Elan-DRC-e (Perkin–Elmer, United States). Blanks, standard and samples were prepared with the same reagents. A multi-elements certified reference solution ICP Standard (Merck) was used for the instrument calibration. Processed blanks were used to calculate the method detection limits (MDL) based on the following equation:

$$\text{MDL} = \text{One-tailed student's } t\text{-test } (p = 0.99\%;\ \text{df} = \text{n} - 1) \times \text{Sr}$$

MDL (mg/kg ww) estimated for each trace elements are the following: As 0.013, Cd 0.002, Co 0.008, Cr 0.003, Cu 0.005, Pb 0.001, Hg 0.0025, Mn 0.005, Ni 0.007, V 0.025, Se 0.03, Sb 0.020 and Zn 0.109.

The quality control was performed with laboratory-fortified matrix (LFM) processed at each batch of digestion, obtaining a recovery ranges from 91.5 to 110% of the nominal concentration.

Statistical analysis was performed with the software SPSS (version 20.0, Inc., IBM, Armonk, NY, USA: IBM Corp.). Results below the MDL were elaborated as MDL/2. The normal distribution was verified using the Kolmogorov–Smirnov test. Since the low number of samples, the Mann-Whitney non-parametric test was used to compare median concentrations between tissues.

2.2. Histological Analysis

Histological analysis was performed according to our standard laboratory procedures [28]. Liver and gills were fixed in 4% formaldehyde (Bio-Optica, Milano, Italy) in PBS buffered to 0.1 M, pH 7.4 (Sigma Life Science) at room temperature for 48 h and processed with Tissue Processing Center TPC 15 Duo (MEDITE®, Burgdorf. Germany). The sections were stained with Haematoxylin-Eosin (HE) (Bio-Optica) and observed under optical microscope (Leica DM750, Monument, CO, USA) equipped with a digital camera (Leica DFC500, Monument, CO, USA).

2.3. Immunohistochemical Analysis

The immunohistochemical protocol was performed on sections to detect mouse monoclonal anti-HSP70 (Gene Tex, 1:1000) and to detect mouse polyclonal anti-MT1 (Abcam, 1:1000); secondary antibody used is FIT-conjugated goat anti-mouse IgG (Sigma-Aldrich, 1:1000). Analysis were performed according to our standard laboratory procedures [29–31]. Slides after mounted with mounting medium containing DAPI (Vectashield, Vector Laboratories, Burlingame, United States), were observed with NIKON ECLIPSE Ci fluorescence microscope and the images taken with the NIKON DS-Qi2 camera.

3. Results and Discussion

The chemical analysis showed that the bioaccumulation of metals in gills was higher in males for Cu ($p < 0.001$), Hg ($p < 0.01$), Se ($p < 0.001$) and Zn ($p < 0.001$), and in females for Cd ($p < 0.001$), Co ($p < 0.001$), Cr ($p < 0.01$), Pb ($p < 0.001$) and V ($p < 0.05$) (Table 1, Figure 1). Nevertheless, the concentrations of all the metals examined were low in both sexes, with higher concentrations of the essential metals Zn and Mn. The analysis of the gastrointestinal content showed a predominance content of Co ($p < 0.05$) and Mn ($p < 0.01$) in males versus an higher concentration of As ($p < 0.01$), Pb ($p < 0.001$), Ni ($p < 0.01$), V ($p < 0.01$) and Se ($p < 0.001$) in females (Table 1, Figure 2). In muscle tissue of both sex, it was observed a lower content of all metals versus the concentrations find in gill and gastrointestinal content. In females, it was found a higher concentration of Cr ($p < 0.01$), Cu ($p < 0.05$), Mn ($p < 0.001$), Ni ($p < 0.001$) and Zn ($p < 0.001$) than males. In males, only the bioaccumulation of Pb ($p < 0.001$) was found higher than females (Table 1, Figure 3).

The results obtained agreed with Bosch et al. [32] which describes a differential bioaccumulation. Particular attention is given to some heavy metals: cadmium, mercury and lead.

The authors hypothesize that the concentrations of heavy metals depend on factors influencing the absorption of metals, such as the simultaneous presence of other xenobiotics, the geographical distribution and the specific biological factors of the species. Overall, the metal concentrations found by us are comparable to those of other studies present in literature [33–35]. These concentrations of metals present in the edible portion of the fish (i.e., muscle) is acceptable as it falls within the limits of international legislation and the specimens analyzed seem to be safe for human consumption. Lead, cadmium and mercury were recorded with values well below the limits imposed by law (0.30 mg/kg for lead, 0.05 mg/kg for cadmium, 0.50 mg/kg for mercury).

Table 1. Descriptive statistics of metals and metalloids (mg/kg ww) in *Merluccius merluccius*.

Tissue	Sex	Statistics	As	Cd	Co	Cr	Cu	Pb	Hg	Mn	Ni	V	Se	Sb	Zn
Gill	Males	Min.	1.115	0.005	0.465	2.468	4.746	2.459	0.030	78.25	1.452	4.598	0.246	0.021	16.43
		Max.	1.924	0.010	0.513	3.215	5.483	5.896	0.058	130.5	3.145	9.452	0.495	0.059	19.75
		Mean	1.536	0.007	0.496	2.851	5.137	3.706	0.043	104.7	2.408	6.501	0.376	0.034	18.76
		S.D.	0.234	0.002	0.017	0.183	0.260	1.032	0.009	18.12	0.559	1.400	0.074	0.012	1.094
	Females	Min.	1.248	0.022	0.745	2.853	3.145	5.424	0.021	78.25	2.139	6.300	0.156	0.021	8.120
		Max.	1.570	0.037	0.985	3.851	4.926	7.952	0.040	120.3	2.770	9.485	0.235	0.062	12.37
		Mean	1.429	0.029	0.880	3.354	4.163	6.486	0.032	97.80	2.517	8.045	0.189	0.044	10.11
		S.D.	0.106	0.005	0.079	0.306	0.483	0.735	0.006	11.53	0.200	1.253	0.024	0.013	1.394
Gastrointestnal content	Males	Min.	0.302	0.001	0.034	0.345	1.952	0.135	0.174	1.324	0.041	0.180	0.141	<0.020	8.125
		Max.	0.601	0.007	0.064	0.699	2.770	0.195	0.251	1.520	0.095	0.264	0.195	<0.020	12.45
		Mean	0.405	0.003	0.048	0.544	2.358	0.165	0.197	1.390	0.066	0.220	0.177	<0.020	9.873
		S.D.	0.084	0.002	0.009	0.107	0.270	0.020	0.023	0.053	0.018	0.028	0.016	/	1.122
	Females	Min.	0.421	0.001	0.021	0.485	1.214	0.214	0.182	1.100	0.065	0.215	0.234	<0.020	8.420
		Max.	0.621	0.009	0.064	0.741	3.254	0.512	0.260	2.164	0.164	0.354	0.485	<0.020	12.48
		Mean	0.505	0.005	0.037	0.620	2.092	0.324	0.210	1.903	0.099	0.280	0.357	<0.020	10.88
		S.D.	0.067	0.003	0.011	0.095	0.585	0.088	0.027	0.348	0.029	0.049	0.087	/	1.331
Muscle	Males	Min.	0.111	0.001	<0.008	0.325	0.218	0.007	0.005	0.230	<0.007	<0.025	0.075	<0.020	0.796
		Max.	0.194	0.009	<0.008	0.852	0.324	0.021	0.016	0.465	<0.007	<0.025	0.164	<0.020	1.500
		Mean	0.151	0.004	<0.008	0.522	0.272	0.014	0.009	0.363	<0.007	<0.025	0.111	<0.020	1.101
		S.D.	0.033	0.002	/	0.146	0.041	0.005	0.003	0.072	/	/	0.023	/	0.213
	Females	Min.	0.114	0.002	<0.008	0.596	0.222	0.001	0.005	1.108	0.011	<0.025	0.085	<0.020	1.745
		Max.	0.384	0.006	<0.008	0.771	0.513	0.009	0.014	1.345	0.035	<0.025	0.254	<0.020	2.224
		Mean	0.212	0.004	<0.008	0.692	0.341	0.003	0.010	1.201	0.021	<0.025	0.137	<0.020	2.072
		S.D.	0.086	0.001	/	0.055	0.079	0.002	0.003	0.078	0.008	/	0.057	/	0.141

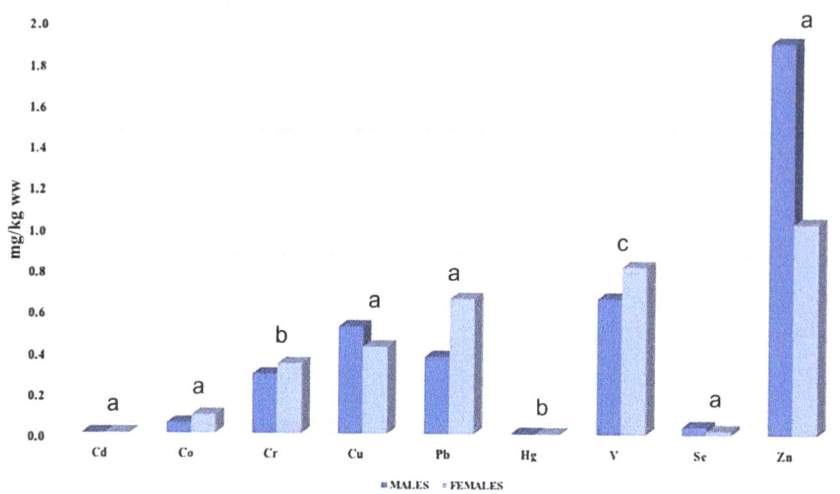

Figure 1. Mean values of metals and metalloids concentrations in gills (mg/kg ww). Significant differences between sex: a, $p < 0.001$; b, $p < 0.01$, c: $p < 0.05$.

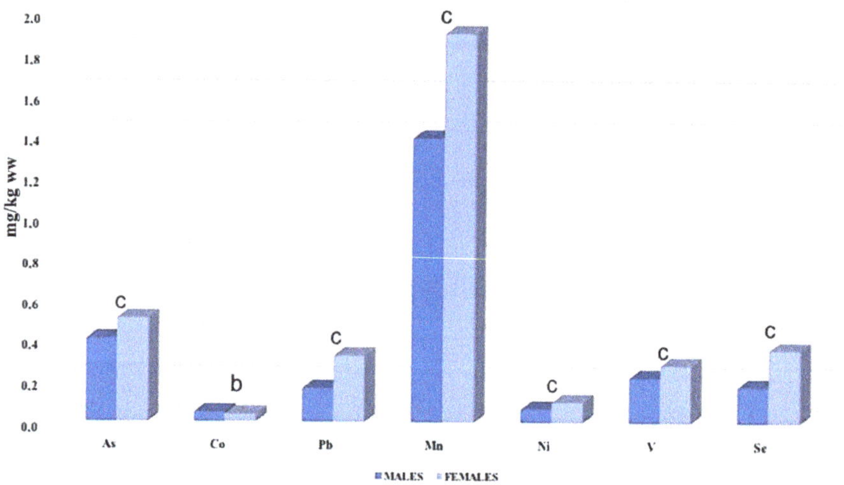

Figure 2. Mean values of metals and metalloids concentrations in gastrointestinal content (mg/kg ww). Significant differences between sex: b, $p < 0.01$, c: $p < 0.05$.

As far as histological investigations are concerned, morphological anomalies of the gill lamellae were not found in the analyzed specimens (Figure 4A,B), but marked liver steatosis was evident in the liver sections (Figure 4C,D). The immunohistochemical investigation revealed a weak expression of MT1 (Figure 5A,C) and HSP70 (Figure 5B,D) in both the target organs (gills and liver) of both sexes.

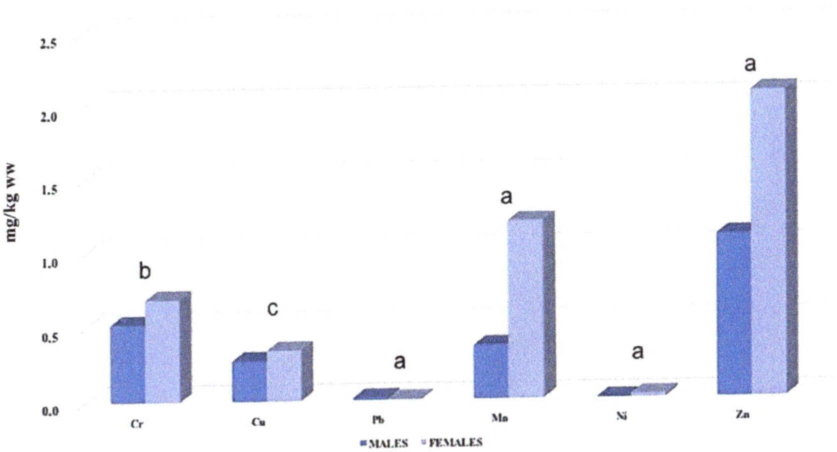

Figure 3. Mean values of metals and metalloids concentrations in muscle (mg/kg ww). Significant differences between sex: a, $p < 0.001$; b, $p < 0.01$, c: $p < 0.05$.

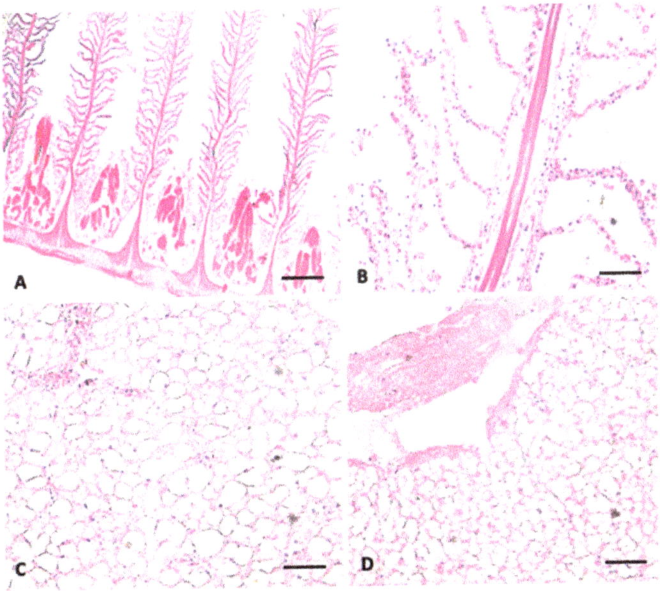

Figure 4. Sections stained with Hematoxylin-Eosin. (**A**,**B**); gills sections (**C**,**D**); male liver sections, is evident a diffuse steatosis. Scale bar A,C and D: 200 µm; B 100 µm.

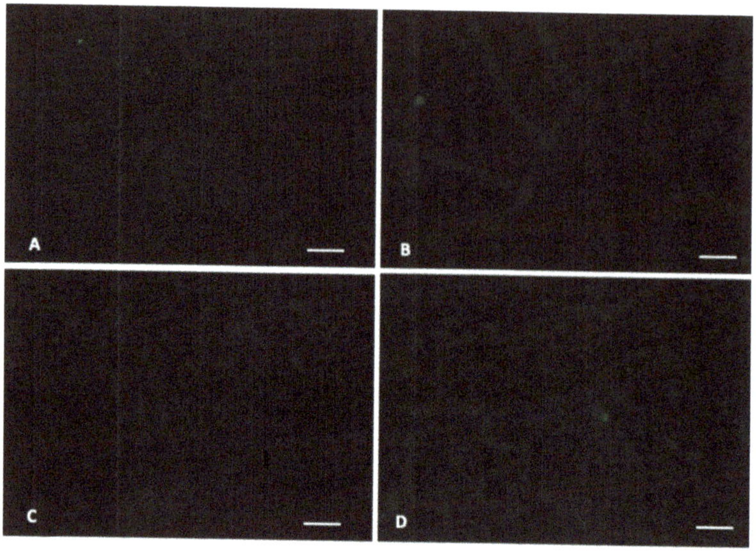

Figure 5. MT 1 expression in male gills (**A**) and liver (**C**). HSP70 expression in male gills (**B**) and in liver (**D**). Using specific antibodies anti-MT1 and anti-HSP70 it has been showed a weak expression of these proteins in cells (green). Scale bar A, C and D: 200 μm; B: 50 μm.

The histological analysis, despite being non-specific about pathogenic noxae, offers the possibility of highlighting any alterations in the gill and liver tissues. The use of these tissue is an efficient biomarker in a preliminary screening for fish and water quality. *Merluccius merluccius*, being a voracious predator, can be considered a useful species for monitoring the environmental situation over time through bioaccumulation and biomagnification studies. Heavy metals contamination in fishery products remains an almost inevitable condition due to the environmental situation facing most of the national fishing areas. Therefore, as already pointed out by European Food Safety Authority (EFSA) [36] in order to reduce this risk to an acceptable level, it is necessary to bridge the lack of information on the conscious consumption of the various fish species and, more generally, it is necessary that consumers can receive a concrete information on the possible risks associated with the consumption of certain species.

In recent decades, with the rapid evolution of molecular, biochemical and pathological technologies, biomarkers have found a very wide application. The toxicity of a contaminant on an organism is usually expressed at a biochemical and molecular level and, because of this, at cellular and tissue levels. As a response to the action of a toxic agent, the body develops adaptive responses that tend to bring the system back into a balanced state. However, when the homeostatic mechanism is not sufficient to balance the action of the toxic agent, the negative effect is manifested at the cellular, tissue and eventually organ level. It can therefore be said that the different responses, homeostatic and otherwise, generated by an organism towards a toxic agent, represent potential markers that can be used in ecotoxicological investigations [37,38].

It can be said that the gills, because of the role of primary importance in interfacing the fish with the surrounding aquatic environment and for their demonstrated reactivity and sensitivity, are optimal candidates for the role of versatile biomarkers in the biomonitoring of aquatic ecosystems. They lend themselves to a study in the field of environmental protection, also aimed at protecting the human being as a user, direct and indirect, of the resources that natural ecosystems can offer. However, also the investigation of ecological indices represents a good method for the monitoring of the environmental quality [39] and these types of studies should be associated to the bioaccumulation studies for a more complete and wide point of view.

4. Conclusions

The diversity of cultures bordering the Mediterranean and the level of anthropization present make interventions aimed at environmental protection difficult. The same international fishing legislation of the Mediterranean Sea is applied with different weights, according to the different social and economic realities on which it acts. To protect the Mediterranean environment, which currently has different pressures at the expense of coastal and marine habitats, it is increasingly necessary, in addition to implementing and applying existing environmental laws, to also adopt integrated approaches based on knowledge of ecosystems in order to better understand their organization and functioning. The result of these negative impacts is an ecosystem in which the survival of fish species is dependent on human intervention. This study analyzes some of the complex variables that currently affect the fish populations of the Mediterranean Sea, intended both as a fish resource of great commercial value, and as a symbol of an ecosystem that now lives below the limits of sustainability. In particular, the contamination level recorded by us in *M. merluccius* in the current study suggested that this important commercial species can be considered sure for human consumption, although we recorded a marked liver steatosis that can be the result of stressing environmental conditions. Further investigation in this direction may help to better understand contamination dynamics and effects on fish population and their potential risks to human health.\

Author Contributions: All authors have made substantial contributions to the conception and the design of the work; have approved the submitted version have agree to be personally accountable for the author's own contributions and for ensuring that questions related to the accuracy or integrity of any part of the work, even ones in which the author was not personally involved, are appropriately investigated, resolved, and documented in the literature. Methodology, A.S., V.S.B., C.C., A.G.; visualization, S.I., G.M., M.F., B.M.L.; writing, review and editing, M.V.B., F.T., R.P., E.M.S., A.S.; conceptualization M.V.B. and F.T. All authors have read and agreed to the published version of the manuscript.

Funding: This study was funded by the Department of Biological, Geological and Environmental Science and by the Department of Medical Sciences, Surgical and Advanced Technologies "G.F. Ingrassia"—Hygiene and Public Health, University of Catania.

Conflicts of Interest: The authors declare no conflict of interest.

References

1. Focardi, S.; Gavilán, J.F.; Barra, R.; Fossi, M.C.; Casini, S.; Salinas, G.; Parra, O. Biochemical biomarkers in fish from different river systems reflect exposure to a variety of anthropogenic stressors. *Bull. Environ. Contam. Toxicol.* **2001**, *66*, 476–483.
2. Di Domenico, A.; Miniero, R. Persistent organic micropollutants in Mediterranean organisms and risk associated. *Ann. Dell'istituto Super. Di Sanita* **2003**, *39*, 1–5.
3. Storelli, M.M.; Marcotrigiano, G.O. Bioindicator organisms: Heavy metal pollution evaluation in the Ionian Sea (Mediterranea Sea–Italy). *Environ. Monit. Assess.* **2005**, *102*, 159–166. [CrossRef] [PubMed]
4. Jaishankar, M.; Tseten, T.; Anbalagan, N.; Mathew, B.B.; Beeregowda, K.N. Toxicity, mechanism and health effects of some heavy metals. *Interdiscip. Toxicol.* **2014**, *7*, 60–72. [CrossRef]
5. Wayland, M. Metals as threats to wildlife. In *Short Course on Wildlife Toxicology*; Canadian Cooperative Wildlife Centre, Western College of Veterinary Medicine, University of Saskatchewan: Saskatoon, SK, Canada, 2000.
6. Riggio, M.; Trinchella, F.; Filosa, S.; Parisi, E.; Scudiero, R. Accumulation of zinc, copper, and metallothionein mRNA in lizard ovary proceeds without a concomitant increase in metallothionein content. *Mol. Reprod. Dev.* **2003**, *66*, 374–382. [CrossRef] [PubMed]
7. Hanas, J.S.; Hazuda, D.J.; Bogenhagen, D.F.; Wu, F.Y.; Wu, C.W. *Xenopus* transcription factor A requires zinc for binding to the 5S RNA gene. *J. Biol. Chem.* **1983**, *258*, 14120–14125.

8. Grummt, I.; Kuhn, A.; Bartsch, I.; Rosenbauer, H. A transcription terminator located upstream of the mouse rDNA initiation site affects rRNA synthesis. *Cell* **1983**, *47*, 901–911. [CrossRef]
9. Dixon, M.; Webb, E.C. Enzyme inhibition and activation. In *Enzymes*; Dixon, M., Webb, E.C., Eds.; Academic Press: New York, NY, USA, 1979; p. 332.
10. Bryce-Smith, D. Zinc deficiency-the neglected factor. *Chem. Br.* **1989**, *25*, 783–786.
11. Cousins, R.J. Absorption, transport, and hepatic metabolism of copper and zinc: Special reference to metallothionein and ruloplasmin. *Physiol. Rev.* **1985**, *2*, 238–309. [CrossRef]
12. Guillou, M.; Ouiniou, F.; Huart, B.; Pagano, G. Comparision of embryonic development and metal contamination in several populations of the sea urchin *Sphaerechinus granularis* (Lamark) exposed to anthropogenic pollution. *Arch. Environ. Contam. Toxicol.* **2000**, *39*, 337–344. [CrossRef]
13. Au, D.W.; Reunov, A.A.; Wu, R.S. Reproductive impairment of sea urchin upon chronic exposure to cadmium. Part II: Effects on sperm development. *Environ. Pollut.* **2001**, *111*, 11–20. [CrossRef]
14. Sunderman, F.W., Jr.; Plowman, M.C.; Kroftova, O.S.; Grbac-Ivankoyic, S.; Foglia, L.; Crivello, J.F. Effects of teratogenic exposures to Zn^{2+}, Cd^{2+}, Ni^{2+}, Co^{2+}, and Cu^{2+} on metallothionein mRNA contents of *Xenopus* embryos. *Pharmcol. Toxicol.* **1995**, *76*, 178–188. [CrossRef] [PubMed]
15. Hanna, L.A.; Peters, J.M.; Wiley, L.M.; Clegg, M.S.; Kenn, C.L. Comparative effects of essential and nonessential metals on preimplantation mouse embryo development in vitro. *Toxicology* **1997**, *116*, 123–131. [CrossRef]
16. Calevro, F.; Beyrsmann, D.; Hartwig, A. Effect of cadmium (II) on the extent of oxidative DNA damage in primary brain cell cultures from *Pleurodeles* larvae. *Toxicol. Lett.* **1998**, *94*, 217–225. [CrossRef]
17. Oskarsson, A.; Palmiger Hallén, I.; Sundberg, J.; Petersson Gravé, K. Risk assessment in relation to neonatal metal exposure. *Analyst* **1998**, *123*, 19–23. [CrossRef] [PubMed]
18. Frenkel, K.; Karkoszka, J.; Cohen, B.; Baranski, B.; Jakubowski, M.; Cosma, G.; Taioli, E.; Toniolo, P. Occupational exposures to Cd, Ni, and Cr modulate titers of antioxidized DNA base autoantibodies. *Environ. Health Perspect.* **1994**, *102*, 221–225.
19. Hartwing, A. Role of DNA repair inhibition in lead and cadmium induced genotoxicity a: Review. *Environ. Health Perspect.* **1994**, *102*, 45–50.
20. Ferrara, F.; Funari, E. *Chemical Risk Associated with the Water Quality of the Adriatic Sea*; ISTISAN 04/4 Reports; Final Report of the Activities Financed by the MURST/CNR "PRISMA2" Project; Istituto Superiore di Sanità: Rome, Italy, 2004.
21. Carrozzi, V.; Di Lorenzo, M.; Massi, D.; Titone, A.; Ardizzone, G.; Colloca, F. Prey preferences and ontogenetic diet shift of European hake *Merluccius merluccius* (Linnaeus, 1758) in the central Mediterranean Sea. *Reg. Stud. Mar. Sci.* **2019**, *25*, 100440. [CrossRef]
22. Martínez-Morcillo, S.; Pérez-López, M.; Míguez, M.P.; Valcárcel, Y.; Soler, F. Comparative study of esterase activities in different tissues of marine fish species *Trachurus trachurus*, *Merluccius merluccius* and *Trisopterus luscus*. *Sci. Total Environ.* **2019**, *679*, 12–22. [CrossRef]
23. Copat, C.; Rizzo, M.; Zuccaro, A.; Grasso, A.; Zuccarello, P.; Fiore, M.; Mancini, G.; Ferrante, M. Metals/Metalloids and Oxidative Status Markers in Saltwater Fish from the Ionic Coast of Sicily, Mediterranean Sea. *Int. J. Environ. Res.* **2020**, *14*, 15–27. [CrossRef]
24. Copat, C.; Grasso, A.; Fiore, M.; Cristaldi, A.; Zuccarello, P.; Signorelli, SS.; Conti, G.O.; Ferrante, M. Trace elements in seafood from the Mediterranean Sea: An exposure risk assessment. *Food Chem. Toxicol.* **2018**, *115*, 13–19. [CrossRef]
25. Ferrante, M.; Pappalardo, A.M.; Ferrito, V.; Pulvirenti, V.; Fruciano, C.; Grasso, A.; Sciacca, S.; Tigano, C.; Copat, C. Bioaccumulation of metals and biomarkers of environmental stress in Parablennius sanguinolentus (Pallas, 1814) sampled along the Italian coast (2017). *Mar. Pollut. Bull.* **2017**, *122*, 288–296. [CrossRef] [PubMed]
26. Tigano, C.; Tomasello, B.; Pulvirenti, V.; Ferrito, V.; Copat, C.; Carpinteri, G.; Mollica, E.; Sciacca, S.; Renis, M. Assessment of environmental stress in Parablennius sanguinolentus (Pallas, 1814) of the Sicilian Ionian coast. *Ecotox. Environ. Saf.* **2009**, *74*, 1278–1286. [CrossRef] [PubMed]
27. Copat, C.; Brundo, M.V.; Arena, G.; Grasso, A.; Oliveri Conti, G.; Ledda, C.; Fallico, F.; Sciacca, S.; Ferrante, M. Seasonal variation of bioaccumulation in *Engraulis encrasicolus* (Linneaus, 1758) and related biomarkers of exposure. *Ecotoxicol. Environ. Saf.* **2012**, *86*, 31–37. [CrossRef]

28. Salvaggio, A.; Tiralongo, F.; Krasakopoulou, E.; Marmara, D.; Giovos, I.; Crupi, R.; Messina, G.; Lombardo, B.M.; Marzullo, A.; Pecoraro, R.; et al. Biomarkers of Exposure to Chemical Contamination in the Commercial Fish Species *Lepidopus caudatus* (Euphrasen, 1788): A Particular Focus on Plastic Additives. *Front. Physiol.* **2019**, *10*, 905. [CrossRef] [PubMed]
29. Brundo, M.V.; Pecoraro, R.; Marino, F.; Salvaggio, A.; Tibullo, D.; Saccone, S.; Bramanti, V.; Buccheri, M.A.; Impellizzeri, G.; Scuderi, V.; et al. Toxicity Evaluation of New Engineered Nanomaterials in Zebrafish. *Front. Physiol.* **2016**, *7*, 130. [CrossRef] [PubMed]
30. D'Errico, G.; Vitiello, G.; De Tommaso, G.; Abdel-Gaward, F.K.; Brundo, M.V.; Ferrante, M.; De Maio, A.; Trocchia, S.; Bianchi, A.R.; Ciarcia, G.; et al. Electron Spin Resonance (ESR) for the study of Reactive Oxygen Species (ROS) on the isolated frog skin (*Pelophylax bergeri*): A non-invasive method for environmental monitoring. *Environ. Res.* **2018**, *165*, 11–18. [CrossRef]
31. Guerriero, G.; Brundo, M.V.; Labar, S.; Bianchi, A.R.; Trocchia, S.; Rabbito, D.; Palumbo, G.; Abdel-Gawad, F.K.; De Maio, A. Frog (*Pelophylax bergeri*, Gunther 1986) endocrine disruption assessment: Characterization and role of skin poly (ADp-ribose) polymerases. *Environ. Sci. Pollut. Res. Int.* **2018**, *25*, 18303–18313. [CrossRef]
32. Bosch, A.C.; O'Neill, B.; Sigge, G.O.; Kerwath, S.E.; Hoffman, L.C. Heavy metals in marine fish meat and consumer health: A review. *J. Sci. Food Agric.* **2016**, *96*, 32–48. [CrossRef]
33. Elnabris, K.J.; Muzyed, S.K.; El-Ashgar, N.M. Heavy metal concentrations in some commercially important fishes and their contribution to heavy metals exposure in Palestinian people of Gaza Strip (Palestine). *J. Assoc. Arab Univ. Basic Appl. Sci.* **2013**, *13*, 44–51. [CrossRef]
34. Gbogbo, F.; Arthur-Yartel, A.; Bondzie, J.A.; Dorleku, W.P.; Dadzie, S.; Kwansa-Bentum, B.; Ewool, J.; Billah, M.K.; Lamtey, A.M. Risk of heavy metal ingestion from the consumption of two commercially valuable species of fish from the fresh and coastal waters of Ghana. *PLoS ONE* **2018**, *13*, e0194682. [CrossRef]
35. Rajeshkumar, S.; Li, X. Bioaccumulation of heavy metals in fish species from the Meiliang Bay, Taihu Lake, China. *Toxicol. Rep.* **2018**, *5*, 288–295. [CrossRef] [PubMed]
36. EFSA. EFSA provides risk assessment on mercury fish: Precautionary advice given to vulnerable groups. *EFSA J.* **2004**, *34*, 1–14.
37. McCarthy, F.; Shugart, L.R. *Biomarkers of Environmental Contamination*; McCarthy, J.F., Shugart, L.R., Eds.; Lewis Publishers: Boca Raton, FL, USA, 1990.
38. Depledge, M.; Fossi, M.C. The role of biomarker in environmental assessment: Invertebrates. *Ecotoxicology* **1994**, *3*, 173–179. [CrossRef]
39. Viana, A.P.; Lucena Frédou, F. Ichthyofauna as bioindicator of environmental quality in an industrial district in the amazon estuary, Brazil. *Braz. J. Biol.* **2014**, *74*, 315–324. [CrossRef] [PubMed]

MDPI
St. Alban-Anlage 66
4052 Basel
Switzerland
Tel. +41 61 683 77 34
Fax +41 61 302 89 18
www.mdpi.com

Journal of Marine Science and Engineering Editorial Office
E-mail: jmse@mdpi.com
www.mdpi.com/journal/jmse